APPLIED LINEAR ALGEBRA

APPLIED LINEAR ALGEBRA

Pramod Gopinath
Abhijith Ajayakumar

KNOWLEDGE BAKERS

ISBN: 978-93-49964-55-6

Published by : **Knowledge Bakers**
Pune, India
info@knowledgebakers.com

PREFACE

Applied Linear Algebra is designed to provide a practical and accessible introduction to the fundamental concepts and techniques of linear algebra, with a strong emphasis on real-world applications. From engineering and computer science to economics and natural sciences, linear algebra plays a pivotal role in solving complex problems involving systems of equations, transformations, and data structures.

This book is intended for undergraduate students, educators, and professionals seeking both a solid theoretical foundation and the ability to apply linear algebraic methods in practical contexts. It bridges the gap between abstract mathematical theory and its implementation in modern applied problems, such as computer graphics, machine learning, optimization, and signal processing.

The chapters progress from basic concepts—such as vectors, matrices, and linear systems—to more advanced topics including eigenvalues, singular value decomposition, and matrix factorizations. Each topic is supplemented with illustrative examples, exercises, and application-driven problems to reinforce understanding and develop computational proficiency.

We have taken care to present material in a clear, logical, and pedagogically sound manner, while also including insights into numerical methods and the use of software tools commonly employed in applied mathematics. Our goal is to empower readers not just to understand linear algebra, but to use it as a powerful tool in analysis and problem-solving.

We are deeply grateful to the academic and professional communities whose research and teaching practices have enriched the development of this book. It is our hope that Applied Linear Algebra will serve as a valuable reference and learning resource for students and practitioners alike, fostering both competence and confidence in one of mathematics' most essential disciplines.

Pramod Gopinath
Abhijith Ajayakumar

CONTENTS

01 MATRICES, MOMENTS AND QUADRATURE: TIME-DEPENDENT PARTIAL DIFFERENTIAL EQUATIONS APPLICATIONS

1. INTRODUCTION: AN OVERVIEW

Consider an autonomous system of ordinary differential equations (ODEs) of the form

$$\mathbf{y}' = F(\mathbf{y}), \quad \mathbf{y}(t_0) = \mathbf{y}_0, \tag{1}$$

such as one that would naturally arise from spatial discretization of a partial differential equation (PDE). Such systems are generally stiff when the underlying PDE has significant diffusive or advective terms, and this stiffness is exacerbated by increasing the spatial resolution. Stiffness leads to significantly increased computational expense for both explicit and implicit time-stepping methods. For explicit methods, the time step is severely restricted by the CFL condition, while implicit methods require the solution of ill-conditioned systems of linear equations during each time step. Such systems are generally sparse and therefore best suited for iterative methods, but for these stiff systems of ODEs, iterative methods require many iterations or a specially developed preconditioner [1].

Exponential propagation iterative (EPI) methods, introduced by Tokman et al. [1, 2], are Runge-Kutta like time-stepping methods that are designed to minimize the number of matrix function-vector products of the form $\mathbf{w} = \varphi(A\tau)\mathbf{b}$, where φ is a smooth function, A is a square matrix, τ is a parameter determined by the time step, and $\mathbf{b}$ is a column vector. The approach used by EPI methods to compute $\mathbf{w}$ for a given symmetric matrix A is to apply the Lanczos algorithm to A with the initial vector $\mathbf{b}$, until we obtain a matrix X_j with orthonormal columns and a tridiagonal matrix T_j such that $X_j^T A X_j = T_j$. Then, we can compute the approximation

$$\mathbf{w}_j = \| \mathbf{b} \|_2 \, X_j \varphi(T_j \tau) \mathbf{e}_1, \tag{2}$$

where $\mathbf{e}_1 = [1 0 \cdots 0]^T$. Since each column $\mathbf{x}_k$ of the matrix X_j is of the form $\mathbf{x}_k = p_{k-1}(A)\mathbf{b}$, where $p_n(A)$ is a polynomial of degree n in A, $\mathbf{w}_j$ is the product of a polynomial in

A of degree $j - 1$ and $\mathbf{b}$. Since the matrix A arises from a stiff PDE, the eigenvalues of A are distributed over several orders of magnitude. As it is generally not possible to accurately approximate $\varphi(\lambda)$ on such a large interval with a low-degree polynomial, a large number of Lanczos iterations is generally necessary in order to obtain a sufficiently accurate of $\mathbf{w}$.

The difficulty that time-stepping methods have with stiffness stems from the fact that low- and high-frequency components of the solution, which change at widely varying speeds, are coupled and therefore must be evolved using a common time step. While the low-frequency components generally make the most significant contribution to the solution, the high-frequency components change most rapidly, and therefore constrain the time step. The greater the spatial resolution, the greater the bandwidth of the solution, thus constraining the time step even further.

This coupling, however, is not the only cause of the greater computational expense incurred by time-stepping methods applied to stiff systems. A key contributing factor is the use of the same polynomial or rational function to approximate all of these components of $\varphi(A\tau)\mathbf{b}$, when such a function cannot effectively approximate $\varphi(\lambda\tau)$ on a large interval except at high degree, which increases computational expense. An alternative is to use Krylov subspace spectral (KSS) methods [3, 4], which employ a component-wise approach to these matrix function-vector products. In KSS methods, which are explicit, each Fourier coefficient of the solution is computed using an interpolating polynomial with frequency-dependent interpolation points. This individualized approach to computing each Fourier component yields high-order accuracy and stability like that of implicit methods, even though KSS methods themselves are explicit.

To date, KSS methods have been applied mostly to linear PDEs on d-dimensional boxes, for $d = 1, 2, 3$, with either periodic or homogeneous Dirichlet or Neumann boundary conditions. A first-order KSS method was applied to nonlinear diffusion equations for image processing by Guidotti et al. [5], but for that problem, a straightforward linearization was used during each time step. In order to compute solutions of nonlinear PDEs with higher-order accuracy in time, it is necessary to treat the nonlinearity more carefully. This can be accomplished by combining KSS methods with another high-order time-stepping method, such as EPI methods.

In this chapter, this combination is presented, for the purpose of solving systems of ODEs of the form (1) that are obtained through spatial discretization of nonlinear PDEs defined on rectangular domains with periodic, homogeneous Dirichlet, or homogeneous Neumann boundary conditions. The proposed algorithm, first described in [6, 7], uses the component-wise approach of KSS methods to more efficiently compute matrix function-vector products of the form $\mathbf{y} = \varphi(A\tau)\mathbf{b}$. The following features distinguish the combined approach from previous work on either EPI or KSS methods:

- Instead of applying Krylov projection (e.g., see [8–10]) with initial vector **b** to compute **y**, it is applied only to a low-frequency projection of **b**, in order to avoid the larger number of iterations that Krylov projection typically incurs at higher spatial resolution. Furthermore, for advection-dominated problems, denoising is applied to the Krylov subspace basis vectors, to eliminate spurious high-frequency oscillations in these vectors which slow convergence.
- For the high-frequency portion of **b**, a KSS method, as described in [11], is used to apply $\varphi(A\tau)$. In this latest version of KSS methods, each Fourier component of the output vector **y** is still approximated using its own block quadrature rule, as before. However, now the quadrature nodes (i.e., frequency-dependent interpolation points for φ) are obtained through high-frequency analysis of block Lanczos iteration, which yields simple formulas for the nodes, instead of having to obtain them by solving an eigenvalue problem [3]. This greatly improves efficiency over the approach used in [3, 4], in which the nodes are obtained by explicitly applying block Lanczos iteration to A for *each* Fourier component, without sacrificing accuracy [11].
- Once the frequency-dependent interpolation points are obtained, it is necessary to construct and apply frequency-dependent interpolating polynomial approximations $p_\omega(\lambda)$ of φ to A, for each wave number ω, and then the Fourier coefficients of the solution are inner products of the form $\hat{e}_\omega^H p_\omega(A\tau)\mathbf{b}$, where $\hat{e}_\omega$ is a discretization of an appropriate Fourier basis function. This paper provides implementation details for this task, and explains how it can be accomplished with as few Fourier transforms as possible [6, 7].

The outline of the paper is as follows. Section 2 gives an overview of KSS methods. Section 3 reviews their acceleration based on asymptotic analysis of recursion coefficients, first presented in [11], and discusses extension to non-self-adjoint operators. Section 4 provides a brief overview of EPI methods and demonstrates the spurious high-frequency oscillations that can occur when using standard Krylov projection within an EPI method. Section 5 describes how KSS and EPI methods are combined. Numerical results are presented in Section 6. Concluding remarks and directions for future work are given in Section7.

2. MATRICES, MOMENTS, QUADRATURE, AND PDE

To provide context for the latest advancements with KSS methods, we begin with an overview of KSS methods as described in [3], applied to the 1-D parabolic PDE

$$u_t + Lu = 0, \quad 0 < x < 2\pi, \quad t > 0, \quad Lu = -(p(x)u_x)_x + q(x)u, \tag{3}$$

$$u(x,0) = f(x), \quad 0 < x < 2\pi, \tag{4}$$

$$u(0,t) = u(2\pi,t), \quad t > 0. \tag{5}$$

where $p(x) > 0$ and $q(x) \geq 0$ on $[0, 2\pi]$, and $q(x)$ is not identically zero. In KSS methods, each Fourier coefficient of the solution $\tilde{u}(x, t_{n+1})$ is obtained by applying the exact solution operator of the PDE to $\tilde{u}(x, t_n)$. For a given wave number ω, such a Fourier coefficient is given by

$$\hat{u}(\omega, t_{n+1}) = \left\langle \frac{1}{\sqrt{2\pi}} e^{i\omega x}, e^{-L\Delta t} \tilde{u}(x, t_n) \right\rangle, \tag{6}$$

where

$$\langle f, g \rangle = \int_0^{2\pi} \overline{f(x)} g(x) dx$$

is the standard inner product on $[0, 2\pi]$ and $e^{-L\Delta t}$ is the solution operator of the PDE (3).

2.1. Matrices, moments, and quadrature

The spatial discretization of (6) yields the bilinear form

$$\mathbf{u}^H \varphi(A) \mathbf{v}, \tag{7}$$

where $\mathbf{u} = \frac{1}{\sqrt{2\pi}} e^{i\omega x}$ and $\mathbf{v} = \tilde{u}(x, t_n)$ are N-vectors, $A = L_N$, where L_N is a spectral discretization of L, and $\varphi(\lambda) = e^{-\lambda t}$.

Because the Sturm-Liouville operator L is self-adjoint and positive definite, it follows that the matrix A is symmetric positive definite. As such, it has real eigenvalues $b = \lambda_1 \geq \lambda_2 \geq \cdots \geq \lambda_N = a > 0$, and corresponding orthonormal eigenvectors $\mathbf{q}_j$, $j = 1, \ldots, N$. From the spectral decomposition of A, we obtain the following representation of (6):

$$\mathbf{u}^H \varphi(A) \mathbf{v} = \sum_{j=1}^{N} \varphi(\lambda_j) \mathbf{u}^H \mathbf{q}_j \mathbf{q}_j^H \mathbf{v}. \tag{8}$$

As described by Golub and Meurant in [12], (8) can be viewed as a Riemann-Stieltjes integral

$$\mathbf{u}^H \varphi(A) \mathbf{v} = \int_a^b \varphi(\lambda) d\alpha(\lambda), \tag{9}$$

where the measure $\alpha(\lambda)$ is defined by

$$\alpha(\lambda)=\begin{cases} 0 & \text{if } \lambda < a \\ \sum_{j=i}^{N} u_j v_j, & \text{if } \lambda_i \le \lambda < \lambda_{i-1}, \quad u_j = \mathbf{u}^H \mathbf{q}_j, \quad v_j = \mathbf{q}_j^H \mathbf{v}. \\ \sum_{j=1}^{N} u_j v_j, & \text{if } b \le \lambda \end{cases}$$

This allows approximation of (6) using Gaussian quadrature rules, where the nodes and weights are obtained by applying the Lanczos algorithm to A with initial vectors **u** and **v** [12].

Figure 1 demonstrates why integrals of the form (8) can be approximated accurately with a small number of nodes in the case where A is a discretization of a differential operator and the vector **u** is used to extract a particular Fourier coefficient of $f(A)\mathbf{v}$. We examine the distribution $d\alpha(\lambda)$ in the case where $\mathbf{u} = \mathbf{v} = e^{i\omega x}$ for small and large values of ω, and for A discretizing a differential operator of the form $-\partial_x a(x)\partial_x$, with $a(x) > 0$ being a smooth function or a piecewise constant function. In either case, $d\alpha(\lambda)$ is mostly concentrated within a portion of the interval of integration $[a, b]$. Gaussian quadrature rules for such integrals naturally target these relevant portions [3, 13].

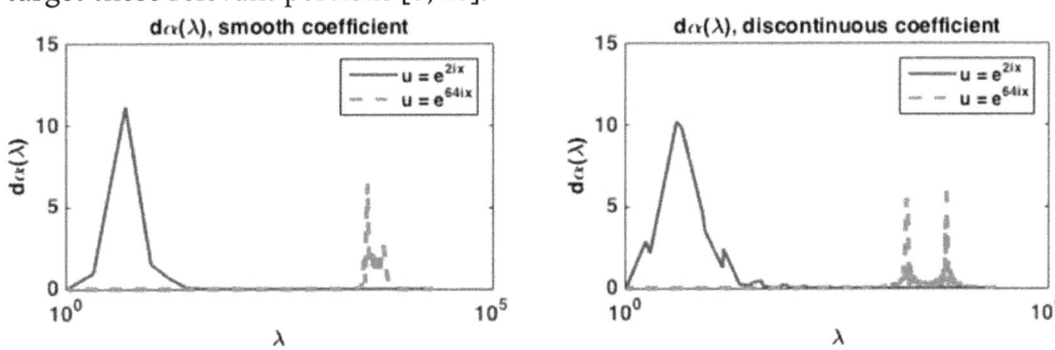

Figure 1. The distribution $d\alpha(\lambda)$ from (8) where the matrix A represents a spectral discretization of a 1-D, second-order differential operator with smooth leading coefficient (top plot) and discontinuous leading coefficient (bottom plot), where u = v is a discretization of e^{2ix} (solid curve) or e^{64ix} (dashed curve).

Block Lanczos algorithm
$X_0 = 0$, $R_0 = [\mathbf{uv}]$, $R_0 = X_1 B_0$ (QR factorization)**for** $n = 1, 2, \ldots, K$
$V = AX_n$
$M_n = X_n^H V$
$R_n = V - X_{n-1}B_{n-1}^H - X_n M_n$
$R_n = X_{n+1} B_n$ (QR factorization)
end

In the case where $\mathbf{u} \neq \mathbf{v}$, the presence of a negative weight would destabilize the quadrature rule [14]. Alternatively, we consider the approximation of the 2 × 2 matrix integral

$$[\mathbf{u} \quad \mathbf{v}]^H \varphi(A)[\mathbf{u} \quad \mathbf{v}]. \qquad (10)$$

We use the most general K-node quadrature formula, as described in [12], to get an approximation for (9) of the form

$$\int_a^b \varphi(\lambda)d\mu(\lambda) = \sum_{j=1}^{2K} \varphi(\lambda_j)\mathbf{v}_j\mathbf{v}_j^H + error, \qquad (11)$$

where, for each j, λ_j is a scalar and $\mathbf{v}_j$ is a 2-vector. Each node λ_j is an eigenvalue of the matrix

$$T_K = \begin{bmatrix} M_1 & B_1^H & & \\ B_1 & M_2 & B_2^H & \\ & \ddots & \ddots & \ddots \\ & & B_{K-1} & M_K \end{bmatrix}, \qquad (12)$$

which is a block-tridiagonal matrix of order $2K$. The vector $\mathbf{v}_j$ consists of the first two elements of the corresponding normalized eigenvector. The matrices M_j and B_j are computed using the block Lanczos algorithm [15], described below.

2.2. Krylov subspace spectral methods

The block KSS method [3, 4] for (3) begins by defining

$$R_0 = \begin{bmatrix} \hat{\mathbf{e}}_\omega & \mathbf{u}^n \end{bmatrix}, \qquad (13)$$

where $\hat{\mathbf{e}}_\omega$ is a discretization of $\frac{1}{\sqrt{2\pi}}e^{i\omega x}$ and $\mathbf{u}^n$ is a discretization of the approximate solution $u(x, t)$ at time $t_n = n\Delta t$. Next, we compute the QR factorization of R_0,

$$R_0 = X_1(\omega)B_0(\omega)$$

which outputs

$$X_1(\omega) = \begin{bmatrix} \hat{\mathbf{e}}_\omega & \dfrac{\mathbf{u}_\omega^n}{\|\mathbf{u}_\omega^n\|_2} \end{bmatrix} \qquad (14)$$

and

$$B_0(\omega) = \begin{bmatrix} 1 & \hat{\mathbf{e}}_\omega^H \mathbf{u}^n \\ 0 & \|\mathbf{u}_\omega^n\|_2 \end{bmatrix}, \qquad (15)$$

where

$$\mathbf{u}_\omega^n = \mathbf{u}^n - \hat{\mathbf{e}}_\omega \hat{\mathbf{e}}_\omega^H \mathbf{u}^n = \mathbf{u}^n - \hat{\mathbf{e}}_\omega \hat{u}(\omega, t_n). \qquad (16)$$

Then we apply the block Lanczos algorithm [15] to the matrix L_N, which comes from the discretization of L, with initial block $X_1(\omega)$. This produces a block tridiagonal matrix T_K of the form (12), where every entry of T_K is a function of ω. Then, at time t_{n+1}, each Fourier coefficient of the solution is

$$[\hat{\mathbf{u}}^{n+1}]_\omega = [B_0(\omega)^H E_{12}^H \exp[-\mathcal{T}_K(\omega)\Delta t] E_{12} B_0(\omega)]_{12}, \quad E_{12} = [\mathbf{e}_1 \quad \mathbf{e}_2]. \quad (17)$$

An inverse FFT applied to the vector of Fourier coefficients $\hat{\mathbf{u}}^{n+1}$ yields the vector $\hat{\mathbf{u}}^{n+1}$, which consists of the values of the solution $u(x, t_{n+1})$ at the grid points.

This algorithm has temporal accuracy $O(\Delta t^{2K-1})$ for parabolic problems [3]. Even higher-order accuracy, $O(\Delta t^{4K-2})$, is obtained for the second-order wave equation [4], due to the second-order time derivative. Furthermore, under appropriate assumptions on the coefficients of the PDE, the 1-node KSS method is unconditionally stable [3, 4]. More generally, the temporal order of accuracy is $O(\Delta t^{(2K-1)d})$, where d is the highest order of a time derivative in the PDE; this order of accuracy has been observed with the Schrödinger equation [16] and Maxwell's equations [18].

3. ASYMPTOTIC ANALYSIS OF BLOCK LANCZOS ITERATION

For each Fourier coefficient of the solution, KSS methods use a different quadrature rule, because the measure $\alpha(\lambda)$ in (8) is defined in terms of the frequency. It follows that if $S(L; \Delta t)$ is the solution operator of the PDE (e.g. $S(L; \Delta t) = e^{-L\Delta t}$ for the PDE (3)), then the function $S(\lambda; \Delta t)$ is approximated by a polynomial of degree $2K$ that interpolates $S(\lambda; \Delta t)$ at different points for each frequency, which are the block Gaussian quadrature nodes λ_j in (11). Taking all Fourier coefficients together, the computed solution at time t_{n+1}, $\mathbf{u}^{n+1}$ can be expressed as

$$\mathbf{u}^{n+1} = \sum_{j=0}^{2K-1} D_j(\Delta t) L_N^j \mathbf{u}^n,$$

where L_N is a discretization of L on an N-point grid and $D_j(\Delta t)$ is a matrix that is diagonal in discrete Fourier space. The diagonal entries are the coefficients of these frequency-dependent interpolating polynomials in power form, with each row corresponding to a different frequency.

In the block KSS method described in [3, 4] and reviewed in the previous section, these interpolation points are the eigenvalues of the block tridiagonal matrix T_K from (12) that is produced by block Lanczos iteration. Although each such matrix is small, computing the eigenvalues and eigenvectors for *each* frequency is computationally expensive. Therefore, in this section, we describe a much faster approach to obtaining estimates of these nodes, at least for high frequencies. This approach was first presented in [11] and generalized in [6, 7]. The basic idea is to examine the entries of T_K as $|\omega| \to \infty$, where ω is the wave number.

3.1. The block case

As in the previous section, let $\mathbf{u}^n$ be a discretization of the approximate solution $u(x, t)$ at time $t_n = n\Delta t$ on a uniform N-point grid. Then, KSS methods use the initial block $R_0=[\hat{\mathbf{e}}_\omega \ \ \mathbf{u}^n]$, for each $\omega = -N/2 + 1, \ldots, N/2$. We start the first iteration of the block Lanczos algorithm by finding the QR-factorization of R_0:

$$R_0 = X_1 B_0,$$

where

$$X_1 = \begin{bmatrix} \hat{\mathbf{e}}_\omega & \dfrac{\mathbf{u}_\omega^n}{\|\mathbf{u}_\omega^n\|_2} \end{bmatrix} \text{ and } B_0 = \begin{bmatrix} 1 & \hat{u}(\omega, t_n) \\ 0 & \|\mathbf{u}_\omega^n\|_2 \end{bmatrix}. \tag{18}$$

with $\mathbf{u}_\omega^n$ is defined as in (16). We note that if the solution u is continuous, then as $|\omega| \to \infty$, $|\hat{u}^n(\omega)| \to 0$, so that in the limit B_0 is diagonal.

The next step is to compute

$$M_1 = X_1^H L_N X_1, \tag{19}$$

where the matrix L_N is a spectral discretization of the operator L defined by $Lu = pu_{xx} + q(x)u$, with p being a constant. Substituting the value of X_1 from (18) into (19) yields

$$M_1 = \begin{bmatrix} \omega^2 p + \bar{q} & \dfrac{\widehat{L_N \mathbf{u}_\omega^n(\omega)}}{\|\mathbf{u}_\omega^n\|_2} \\ \dfrac{\overline{\widehat{L_N \mathbf{u}_\omega^n(\omega)}}}{\|\mathbf{u}_\omega^n\|_2} & R(L_N, \mathbf{u}_\omega^n) \end{bmatrix},$$

where $\bar{q}$ is the mean of $q(x)$ on $(0, 2\pi)$, $\widehat{L_N \mathbf{u}_\omega^n(\omega)} = \hat{\mathbf{e}}_\omega^H L_N \mathbf{u}_\omega^n$ is the Fourier coefficient of the grid function $L_N \mathbf{u}^n$ corresponding to the wave number ω, and $R(L_N, \mathbf{u}_\omega^n) = \dfrac{\langle \mathbf{u}_\omega^n, L_N \mathbf{u}_\omega^n \rangle}{\langle \mathbf{u}_\omega^n, \mathbf{u}_\omega^n \rangle}$ is the Rayleigh quotient of L_N and $\mathbf{u}_\omega^n$. As $|\omega|$ increases, the Fourier coefficients of a continuous function decay to zero; therefore, as long as the solution is sufficiently regular, the non-diagonal entries of M_1 become negligible, that is,

$$M_1 \approx \begin{bmatrix} \omega^2 p + \bar{q} & 0 \\ 0 & R(L_N, \mathbf{u}^n) \end{bmatrix}.$$

Proceeding with the iteration, and neglecting any terms that are Fourier coefficients or are of lower order in ω, we obtain

$$R_1 = L_N X_1 - X_1 M_1 \approx \left[\tilde{\mathbf{q}} \hat{\mathbf{e}}_\omega \quad \frac{L_N \mathbf{u}^n_\omega}{\| \mathbf{u}^n_\omega \|_2} - R(L_N, \mathbf{u}^n_\omega) \frac{\mathbf{u}^n_\omega}{\| \mathbf{u}^n_\omega \|_2} \right],$$

where $\mathbf{q}$ is a vector consisting of the value of $q(x)$ at the grid points, $\tilde{q}=q-\bar{q}$, and multiplication of vectors is component-wise.

To obtain X_2, we perform the QR-factorization $R_1 = X_2 B_1$. We note that the (1, 2) entry of B_1, modulo lower-order terms, is the Fourier coefficient $\hat{v}_1(\omega)$,, where

$$\mathbf{v}_1 = \tilde{\mathbf{q}} \left(\frac{L_N \mathbf{u}^n_\omega}{\| \mathbf{u}^n_\omega \|_2} - R(L_N, \mathbf{u}^n_\omega) \frac{\mathbf{u}^n_\omega}{\| \mathbf{u}^n_\omega \|_2} \right).$$

It follows that if the coefficient $q(x)$ and solution $u(x, t)$ are sufficiently regular, then B_1 approaches a diagonal matrix as $|\omega| \to \infty$, just as B_0 does. Continuing this process, it can be shown that given sufficient regularity of the solution and coefficients of L, each block M_j or B_j of T^K from (12) becomes approximately diagonal at high frequencies.

Because $[T_K]ij \approx 0$ when $i+j$ is odd in the high-frequency case, it follows that if the rows and columns of T_K are permuted so that odd-numbered and even-numbered rows and columns are grouped together and in order, then the eigenvalue problem for T_K approximately *decouples*. Therefore, approximate eigenvalues of T_K can be obtained by computing the eigenvalues of the tridiagonal matrices obtained by performing "non-block" Lanczos on L_N with initial vectors equal to the two columns of R_0 separately, rather than applying block Lanczos with these two columns together in the initial block. In [11], it is shown that this decoupling also takes place if the leading coefficient $p(x)$ of L is *not* constant.

3.2. The non-block case

The decoupling observed in the preceding discussion reveals that we can obtain approximations of half of the block Gaussian quadrature nodes for all Fourier coefficients by applying "non-block" Lanczos iteration to the matrix L_N with initial vector $\mathbf{u}^n$, the computed solution, as is done in Krylov projection methods such as those described in [8–10]. These nodes, denoted by $\lambda_1, \ldots, \lambda_m$ where m is the number of iterations, will be referred to as *frequency-independent nodes.* Because this iteration does not depend on the wave number ω, the frequency-independent nodes need only be computed once for each vector $\mathbf{u}$ for which an expression of the form $\varphi(L_N \tau)\mathbf{u}^n$ is required. The other half of the nodes can be estimated through an asymptotic analysis of Lanczos iteration applied to L_N with initial vector $\hat{\mathbf{e}}_\omega$ [11]; these are called *frequency-dependent nodes* and will be denoted by $\lambda_{1,\omega}, \ldots, \lambda_{m,\omega}$.

As an example, we consider the case where the matrix A comes from a spectral discretization of the operator $Lu = -pu_{xx} + q(x)u$, where p is a constant, and assuming periodic boundary conditions [11]. Carrying out three iterations, which corresponds to a fifth-order accurate KSS method for a parabolic PDE, yields the following recursion coefficients as functions of the wave number ω, after neglecting lower-order terms:

$$\begin{bmatrix} \alpha_1 & \overline{\beta_1} & 0 \\ \beta_1 & \alpha_2 & \overline{\beta_2} \\ 0 & \beta_2 & \alpha_3 \end{bmatrix} \approx \begin{bmatrix} p\omega^2 & \|\tilde{\mathbf{q}}\|_2 & 0 \\ \|\tilde{\mathbf{q}}\|_2 & p\omega^2 & 2p|\omega|\|\mathbf{q}_x\|_2/\|\tilde{\mathbf{q}}\|_2 \\ 0 & 2p|\omega|\|\mathbf{q}_x\|_2/\|\tilde{\mathbf{q}}\|_2 & p\omega^2 \end{bmatrix}.$$

It follows that the frequency-dependent nodes can easily be estimated as

$$\lambda_{1,\omega} = p\omega^2, \quad \lambda_{i,\omega} = p\omega^2 \pm \sqrt{\beta_1^2 + \beta_2^2}, \quad i = 2,3 \tag{20}$$

whereas for a third-order KSS method, the frequency-dependent nodes are

$$\lambda_{1,\omega} = p\omega^2 + \beta_1, \quad \lambda_{2,\omega} = p\omega^2 - \beta_1.$$

In [6, 7], similar formulas for the nodes are derived for a PDE with homogeneous Neumann boundary conditions, and for a 2-D PDE with periodic boundary conditions.

When the matrix A is a finite-difference representation of the underlying differential operator, the block Gaussian quadrature nodes can be represented more accurately if formulas for the eigenvalues of symmetric Toeplitz matrices are used for the leading-order terms in the nodes. For example, in (20), $p\omega^2$ is replaced by $2p(N/\pi)^2(1 - \cos(\pi\omega/N))$, where N is the number of grid points.

3.3. Non-self-adjoint operators

The theory developed in [12] applies to symmetric positive definite matrices, but this property is not essential [17, 18]. That said, care must be taken with nonsymmetric matrices, as a straightforward use of unsymmetric Lanczos to treat bilinear forms as Riemann-Stiejtjes integrals, as described in [19], can suffer from "serious breakdown" [20]. Therefore, a more robust approach for applying KSS methods to linear PDE with non-self-adjoint spatial differential operators is to use Arnoldi iteration to approximate $\varphi(A\tau)\mathbf{b}$. The algorithm for Arnoldi iteration, applied to a matrix A and initial vector $\mathbf{z}_0$, is as follows:

Arnoldi iteration

$v_1 = \mathbf{z}_0/\|\mathbf{z}_0\|_2$
for $j = 1, 2, \ldots$
  $\mathbf{z}_j = A\mathbf{v}_j$
  for $k = 1, 2, \ldots, j$
    $h_{kj} = \mathbf{v}_k^H \mathbf{z}_j$
    $\mathbf{z}_j = \mathbf{z}_j - h_{kj}\mathbf{v}_k$
  end
  $h_{j+1,j} = \|\mathbf{z}_j\|_2$
  $\mathbf{v}_{j+1} = \mathbf{z}_j/h_{j+1,j}$
end

The output of Arnoldi iteration is an upper Hessenberg matrix H_m, and a matrix V_m with orthonormal columns, such that

$$AV_m = V_m H_m + h_{m+1,m}\mathbf{z}_{m+1}\mathbf{e}_m^H. \tag{21}$$

By analogy with (17), to approximate $\mathbf{u}^{n+1} = \varphi(A\tau)\mathbf{u}^n$, we can compute each Fourier coefficient $[\hat{\mathbf{u}}^{n+1}]_\omega$ of $\mathbf{u}^{n+1}$ by applying *block* Arnoldi iteration [21] to A, with initial block $R_0(\omega)$ as defined in (13). After m iterations, this iteration produces a block upper Hessenberg matrix $H_m(\omega)$, from which we obtain

$$[\hat{\mathbf{u}}^{n+1}]_\omega = \left[B_0(\omega)^H E_{12}^H \varphi(H_m(\omega)\tau)E_{12}B_0(\omega)\right]_{12}, \tag{22}$$

where $B_0(\omega)$ and E_{12} are as defined in (15) and (17), respectively. As shown in [6, 7], like with block Lanczos, the eigenvalue problem for $H_m(\omega)$ approximately decouples for high frequencies, due to the decay of the Fourier coefficients of $\mathbf{u}^n$. It follows that we can easily estimate the frequency-dependent eigenvalues of $H_m(\omega)$, which are used as interpolation points for a polynomial approximation of $\varphi(\lambda)$, by applying "non-block" Arnoldi iteration, as described in the above algorithm, with an initial vector $\mathbf{z}_0$ chosen to be a discretization of an appropriate Fourier basis function. Additional details can be found in [6, 7].

4. EPI METHODS

In this section, we give an overview of EPI methods, as developed by Tokman et al. in [1, 2]. To solve an autonomous system of ODEs of the form (1), a time step from time t_n to time $t_{n+1} = t_n + \Delta t$ is taken as follows. First, the time derivative $F(\mathbf{y})$ is expressed in terms of its Taylor expansion around $\mathbf{y}(t_n)$, which yields

$$\frac{d\mathbf{y}}{dt} = F(\mathbf{y}(t_n)) + A_n(\mathbf{y}(t) - \mathbf{y}(t_n)) + R(\mathbf{y}(t)), \tag{23}$$

where A_n is the Jacobian matrix of $\mathbf{F}(\mathbf{y})$ evaluated at $\mathbf{y}(t_n)$, and $R(\mathbf{y}(t))$ is the Taylor remainder. Next, we multiply both sides by an integrating factor $e^{-A_n t}$ and then integrate from t_n to t_{n+1} to obtain

$$\mathbf{y}(t_{n+1}) = \mathbf{y}(t_n) + [e^{A_n \Delta t} - I]A_n^{-1}F(\mathbf{y}(t_n)) + \int_{t_n}^{t_{n+1}} e^{A_n(t_{n+1}-\tau)} R(\mathbf{y}(\tau))d\tau. \qquad (24)$$

The integral on the right side is then approximated using a quadrature rule. This requires the evaluation of matrix function-vector products of the form $\varphi(A\tau)\mathbf{b}$, with $A = A_n$, for various "φ-functions" and various choices of the vectors **b** and scaling factors τ, which are determined so as to satisfy order conditions.

Any such matrix function-vector product is computed using Krylov projection. Arnoldi iteration is applied to A (or Lanczos iteration, if A is symmetric) with initial vector **b**. After m iterations, we obtain (21), from which we obtain an approximation

$$\varphi(A\tau)\mathbf{b} \approx \|\mathbf{b}\|_2 V_m \varphi(H_m \tau)\mathbf{e}_1, \qquad (24)$$

where, as before, H_m is upper Hessenberg (or tridiagonal of A is symmetric), and the columns of V_m form an orthonormal basis for the Krylov subspace

$$\mathcal{K}(A, \mathbf{b}, m) = \operatorname{span}\{\mathbf{b}, Ab, A^2\mathbf{b}, \ldots, A^{m-1}\mathbf{b}\}.$$

The accuracy of this approximation is discussed in [9]. In the case where A is ill-conditioned, the number of iterations m needed for convergence of (25) can be quite large, and this is exacerbated by increasing the spatial resolution in the discretization of the underlying PDE from which (1) arises.

When the number of iterations is large, an additional issue, particularly for advection-dominated problems, is the appearance of spurious high-frequency oscillations in the columns of V_m, even if the initial vector **b** represents a smooth function. As can be seen in Figure 2, even after relatively few iterations, these oscillations can occur and reduce the columns of V_m to essentially noise.

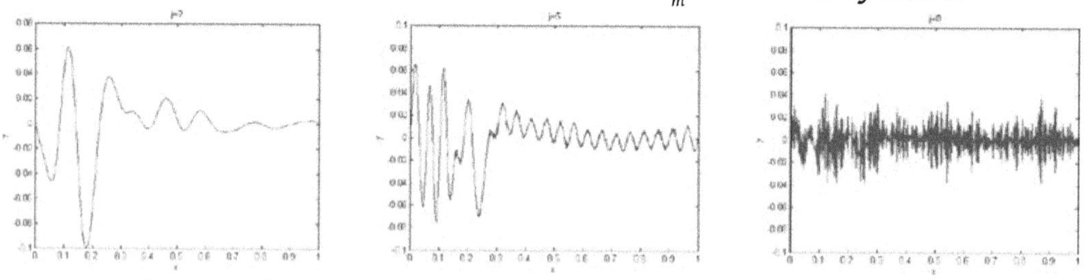

Figure 2. Columns of V_m from (25) generated by Arnoldi iteration applied to the matrix from Burgers' equation (see Section 6.2).

This can be alleviated by filtering out high-frequency components of the columns of V_m after each matrix-vector multiplication. As can be seen in Figure 3, the spurious high-frequency oscillations are reduced, resulting in more meaningful data in V_m. Future work will include the automatic, adaptive selection of an appropriate threshold for filtering high-frequency components.

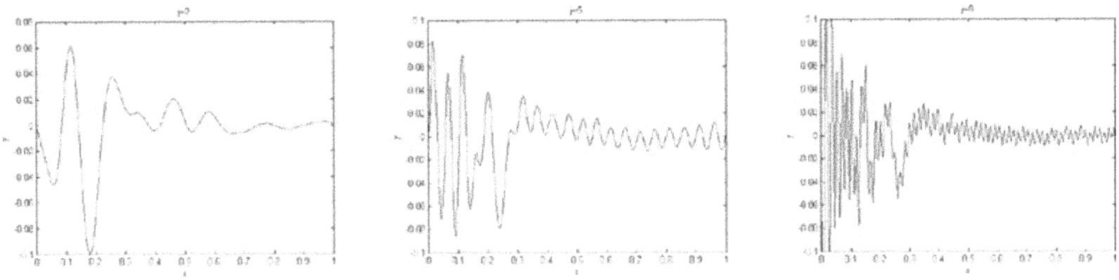

Figure 3. Columns of V_m from (25) generated by Arnoldi iteration, with denoising, applied to the matrix from Burgers' equation (see Section 6.2).

The behavior of the unfiltered Krylov vectors is not surprising, as similar behavior is displayed by the unsmoothed Fourier method applied to hyperbolic PDEs [22]. In that work, the proposed remedies were to either use filtering, or increase the number of grid points; the former remedy serves as the motivation for denoising in this context. It is worth noting that in both the unfiltered and filtered cases, no loss of orthogonality was observed in the Krylov vectors; that is, $1/\sqrt{m}\,\| V_m^T V_m - I_m \|_F$ was negligibly small (that is, on the order of 10^{-10} or smaller) in both cases.

5. KSS-EPI METHODS

We now describe the combination of KSS and EPI methods. The EPI method itself is not changed; what is modified is the approach to computing any matrix function-vector product of the form $\varphi(A\tau)\mathbf{b}$. The main steps in the new approach are as follows:

- First, it is necessary to determine a cutoff frequency N_c. In the numerical experiments presented in this paper, the value of N_c has been determined by experimentation; it is demonstrated in [7] that the performance is not unduly sensitive to the choice of N_c. In future work, an adaptive approach to choosing N_c will be developed.
- Using an FFT, $\mathbf{b}$ is decomposed into low-frequency and high-frequency components $\mathbf{b}_L$ and $\mathbf{b}_H$. Specifically, $\mathbf{b} = \mathbf{b}_L + \mathbf{b}_H$, where each Fourier coefficient of $\mathbf{b}_L$ with wave number $\vec{\omega}$ is zero if $\|\vec{\omega}\|_\infty \geq N_c$.
- $\varphi(A\tau)\mathbf{b}_L$ is computed using Krylov projection, as is normally done in EPI methods.
- $\varphi(A\tau)\mathbf{b}_H$ is computed using a KSS method, as described in Section 3.
- The results of the preceding two steps are added to obtain an approximation of $\varphi(A\tau)\mathbf{b}$.

We now provide details on how step 4 can be performed efficiently, by minimizing the number of FFTs. To simplify the exposition, we consider the 1-D case, with periodic boundary conditions. For each wave number ω, we obtain the frequency-dependent nodes $\{\lambda_{1,\omega}, \lambda_{2,\omega}, ..., \lambda_{K,\omega}\}$, as described in Section 3, by approximating the entries of the tridiagonal matrix $T_K(\omega)$ produced by Lanczos iteration with initial vector $\hat{\mathbf{e}}_\omega$. As demonstrated in Section 3, this is readily accomplished using the coefficients of the spatial differential operator. Next, we use Krylov projection with initial vector $\mathbf{b}_H$, as in Section 4, and compute the eigenvalues of the resulting tridiagonal (or Hessenberg, if A is nonsymmetric) matrix. These are the frequency-independent nodes $\{\lambda_1, \lambda_2, ..., \lambda_K\}$.

The Fourier coefficient of $\varphi(A\tau)\mathbf{b}_H$ corresponding to the wave number ω is obtained by computing the same Fourier coefficient of $p_{2K-1,\omega}(A\tau)\mathbf{b}_H$, where $p_{2K-1,\omega}$ is the polynomial interpolant of $\varphi(\lambda)$ with interpolation points $\{\lambda_i, \lambda_{i,\omega}\}_{i=1}^K$. Expressing this interpolant in Newton form, we have

$$p_{2K-1,\omega}(\lambda) = \sum_{j=1}^{K} \varphi[\lambda_1, \ldots, \lambda_j] \prod_{i=1}^{j-1} (\lambda - \lambda_i) + \sum_{j=1}^{K} \varphi[\lambda_1, \ldots, \lambda_K, \lambda_{1,\omega}, \ldots, \lambda_{j,\omega}] \left(\prod_{i=1}^{j-1} (\lambda - \lambda_{i,\omega}) \right) \prod_{k=1}^{K} (\lambda - \lambda_k). \quad (26)$$

Arranging the interpolation points in the order indicated above allows us to reduce the number of FFTs needed. Using the relation from Lanczos iteration,

$$AX_K = X_K T_K + \mathbf{r}_K \mathbf{e}_K^T, \quad (27)$$

we define

$$\begin{aligned}
\mathbf{v} = p_{K-1}(A)\mathbf{b}_H &= \{\varphi[\lambda_1] + \varphi[\lambda_1, \lambda_2](A - \lambda_1 I) + \cdots \\
&\quad + \varphi[\lambda_1, \lambda_2, \ldots, \lambda_K](A - \lambda_1 I) \cdots (A - \lambda_{K-1} I)\} \mathbf{b}_H \\
&= \| \mathbf{b}_H \|_2 X_K p_{K-1}(A) \mathbf{e}_1, \\
\mathbf{w} = q_K(A)\mathbf{b}_H &= (A - \lambda_1 I) \cdots (A - \lambda_K I) \mathbf{b}_H \\
&= \beta_1 \beta_2 \ldots \beta_{K-1} \mathbf{r}_K \\
&= \beta_1 \ldots \beta_K X_{K+1} \mathbf{e}_{K+1},
\end{aligned}$$

and

$$\begin{aligned}
\tilde{p}_{K-1,\omega}(\lambda) &= \varphi[\lambda_1, \ldots, \lambda_K, \lambda_{1,\omega}] + \varphi[\lambda_1, \ldots, \lambda_{2,\omega}](\lambda - \lambda_{1,\omega}) + \cdots \\
&\quad + \varphi[\lambda_1, \ldots, \lambda_{K,\omega}](\lambda - \lambda_{1,\omega}) \cdots (\lambda - \lambda_{K-1,\omega}) \\
&= C_{K-1}^{\omega} \lambda^{K-1} + \cdots + C_1^{\omega} \lambda + C_0^{\omega},
\end{aligned}$$

where C_j^{ω},, for $j = 0, 1, ..., K - 1$, are the coefficients of $\tilde{p}_{K-1,\omega}$ i in power form, which can easily be computed by repeated application of nested multiplication to the last K terms of the Newton form of $p_{2K-1,\omega}$.

Finally, using F to denote the discrete Fourier transform, we have

$$\varphi(A\tau)\mathbf{b}_H \approx p_{2K-1,\omega}(A\tau)\mathbf{b}_H = \mathbf{v} + \tilde{p}_{K-1,\omega}(A)\mathbf{w} = v + \mathsf{F}^{-1}\sum_{j=0}^{K-1}[C_j^{\omega}]\mathcal{F}A^j\mathbf{w}, \tag{28}$$

and it can easily be seen that the solution at each time step requires K FFTs and one inverse FFT. It is worth noting that once the Krylov subspace vectors $A^j\,\mathbf{w}, j = 0, 1, \ldots, K - 1$ are computed, the K FFTs can be performed in parallel.

6. NUMERICAL RESULTS

In this section, we compare several versions of EPI methods, as applied to two test problems; additional test problems are featured in [6, 7]. The versions differ in the way in which they compute matrix function-vector products of the form $\varphi(A\tau)\mathbf{b}$:

- Standard Krylov projection, as in (25), hereafter referred to as "Krylov-EPI", either with or without denoising as in Section 4;
- Using the KSS approach, as described in Section 5, hereafter referred to as "KSS-EPI";
- Newton interpolation using Leja points [23], hereafter referred to as "LEJA"; and
- Adaptive Krylov projection [24], hereafter referred to as "AKP".

All of these approaches are used in the context of two EPI methods. The first is a third-order, two-stage EPI method [1]

$$\begin{aligned} Y_1 &= \mathbf{y}_n + \frac{1}{3}ha_{11}\varphi_1\left(\frac{1}{3}hA\right)F(\mathbf{y}_n), \\ \mathbf{y}_{n+1} &= \mathbf{y}_n + h\varphi_1(hA)F(\mathbf{y}_n) + 3hb_1\varphi_2(hA)[F(Y_1) - F(\mathbf{y}_n) - A(Y_1 - \mathbf{y}_n)], \end{aligned} \tag{29}$$

where $a_{11} = 9/4$ and $b_1 = 32/81$, and

$$R(Y_1) = F(Y_1) - F(\mathbf{y}_n) - A(Y_1 - \mathbf{y}_n).$$

For this method,

$$\varphi_1(\lambda) = \frac{e^{\lambda} - 1}{\lambda}, \quad \varphi_2(\lambda) = \frac{e^{\lambda} - \lambda - 1}{\lambda^2}, \quad \varphi_3(\lambda) = \frac{e^{\lambda}(6-\lambda) - (6 + 5\lambda + 2\lambda^2)}{\lambda^3}.$$

The second is a fifth-order, three-stage EPI method [25]

$$\begin{aligned} Y_1 &= \mathbf{y}_n + ha_{11}\psi_1(g_{11}hA)F(\mathbf{y}_n), \\ Y_2 &= \mathbf{y}_n + ha_{21}\psi_1(g_{21}hA)F(\mathbf{y}_n) + ha_{22}\psi_2(g_{22}hA)R(Y_1), \\ \mathbf{y}_{n+1} &= \mathbf{y}_n + hb_1\psi_1(g_{31}hA)F(\mathbf{y}_n) + hb_2\psi_2(g_{32}hA)R(Y_1) + \\ &\quad hb_3\psi_3(g_{33}hA)[-2R(Y_1) + R(Y_2)], \end{aligned} \tag{30}$$

where

$$\psi_i(z) = \sum_{j=1}^{i} p_{ij}\varphi_j(z), \quad i = 1,2,3,$$

and the coefficients g_{ij}, a_{ij}, b_j, and p_{ij} are obtained from the description of the EPIRK5s3 method in [25].

For all methods used to compute matrix function-vector products, efficiency will be measured in two ways: (1) total execution time required to integrate over the entire time interval, and (2) the average number of matrix-vector products (plus the average number of FFTs, in the case of KSS-EPI) performed for each evaluation of a matrix function-vector product of the form $\varphi(A\tau)\mathbf{b}$. The phrase "number of iterations" is used throughout this section to refer to this quantity. Throughout this section, for the purpose of discussing performance as a function of spatial resolution, N refers to the number of grid points per dimension.

6.1. Diffusive problem

For our first test problem, we choose a diffusion-dominated PDE: the 2-D Allen-Cahn equation,

$$u_t = \alpha \nabla^2 u + u - u^3, \quad x, y \in [0,1], \quad t \in [0, 0.2]$$

with $\alpha = 0.1$. We impose homogeneous Neumann boundary conditions, and the initial condition

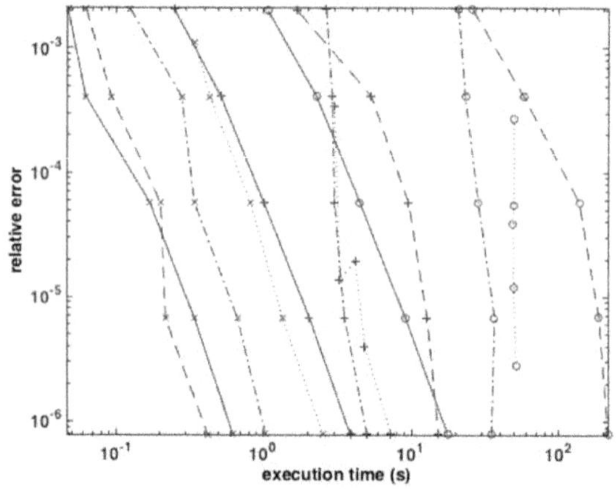

Figure 4. Relative error plotted against execution time for solving the Allen-Cahn equation (31) using the third-order EPI method (29). Matrix function-vector products are computed using KSS-EPI with denoising (solid curves), Krylov-EPI (dashed curves), AKP (dashed-dotted curves), and LEJA (dotted curves), on grids with N = 50 ('+' markers), 150 ('x' markers) and 300 ('o' markers) points per dimension. Time steps used are $\Delta t = (0.2)2^{-p}$, for p = 0, 1, 2, 3, 4.

$$u_0(x, y) = 0.4 + 0.1\cos(2\pi x)\cos(2\pi y).$$

The Laplacian is discretized using a centered finite difference. For KSS-EPI, the low-frequency portion $\mathbf{b}_L$ consists of all components with wave numbers $\omega_i \leq 7$, $i = 1, 2$. That is, for this problem, the value of N_c, as defined in the previous section, is 7. The low value of this threshold is due to the smoothness of the initial data.

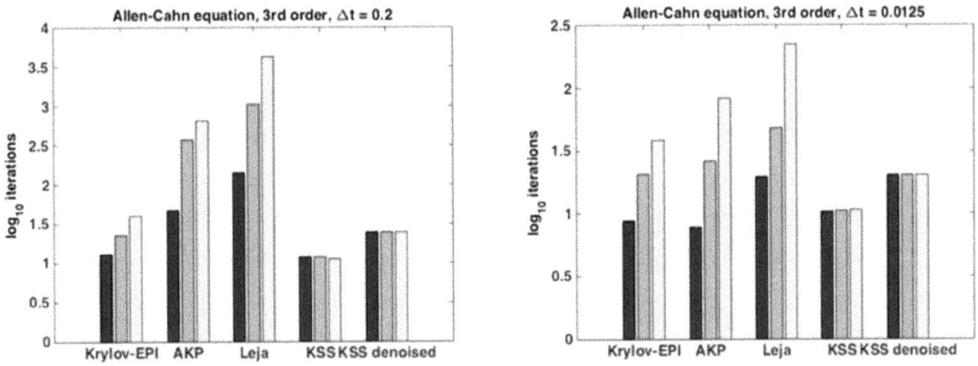

Figure 5. Average number of matrix-vector products, shown on a logarithmic scale, per matrix function-vector product evaluation for each method when solving the Allen-Cahn equation (31) using the third-order EPI method (29). For KSS and KSS denoised, FFTs are also included. For each method, bars correspond to grid sizes of N = 50, 150, 300 points per dimension, from left to right. Left plot: $\Delta t = 0.2$. Right plot: $\Delta t = 0.0125$.

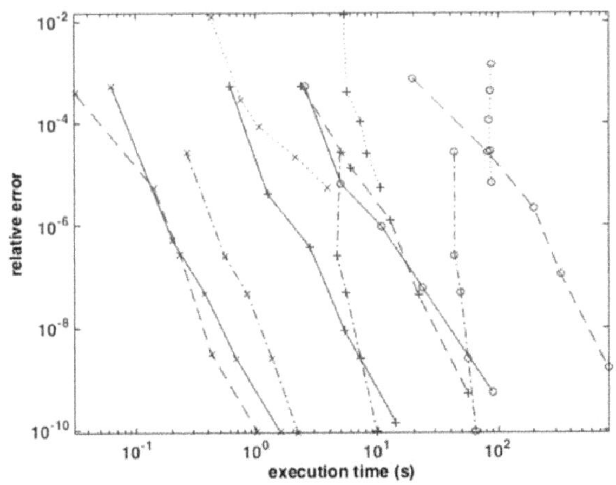

Figure 6. Relative error plotted against execution time for solving the Allen-Cahn equation (31) using the fifth-order EPI method (30). Matrix function-vector products are computed using KSS-EPI with denoising (solid curves), Krylov-EPI (dashed curves), AKP (dashed-dotted curves), and LEJA (dotted curves), on grids with N = 50 ('+' markers), 150 ('x' markers), and 300 ('o' markers) points per dimension. Time steps used are $\Delta t = (0.2)2^{-p}$, for p = 0, 1, 2, 3, 4.

The results are shown in Figures 4 and 5 for the third-order EPI method (29), and in Figures 6 and 7 for the firth-order EPI method (30). It can be seen from Figures 4 and 6 that for both EPI methods, the accuracy of all four approaches to computing matrix function-vector products are comparable. However, the efficiency and scalability of these four approaches are very different. In particular, these figures show that for KSS-EPI methods, the growth in execution time as a function of the number of grid points per dimension is much less. For example, note that for the coarsest grid, with $N = 50$, the speed of KSS-EPI is similar to that of Krylov-EPI, but for $N = 150$ and $N = 300$, KSS-EPI is much faster. Furthermore, as N increases, this gap in performance grows. This is due to the fact that while both methods use Krylov projection, KSS-EPI applies it to an initial vector with only low-frequency components, thus significantly reducing the number of iterations needed for convergence.

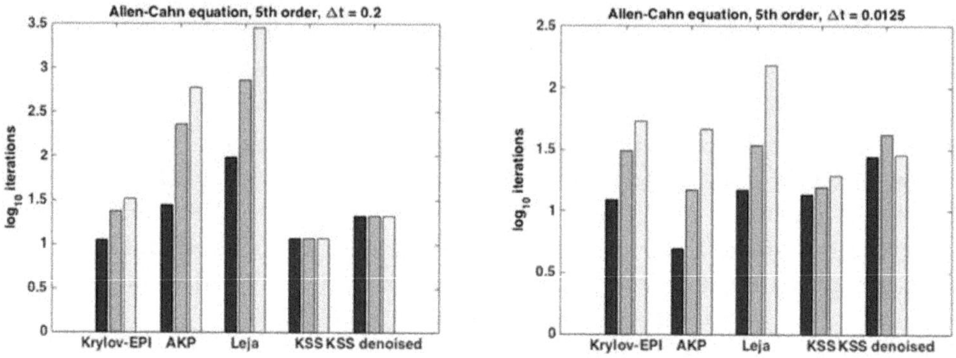

Figure 7. Average number of matrix-vector products, shown on a logarithmic scale, per matrix function-vector product evaluation for each method when solving the Allen-Cahn equation (31) using the fifth-order EPI method (29). For KSS and KSS denoised, FFTs are also included. For each method, bars correspond to grid sizes of N = 50, 150, 300 points per dimension, from left to right. Left plot: Δt = 0.2. Right plot: Δt = 0.0125.

The difference in scalability is more clearly illustrated in Figures 5 and 7. It can be seen that for KSS-EPI, the number of overall iterations (matrix-vector multiplications + FFTs) shows almost no sensitivity to the grid size, compared to Krylov-EPI, AKP and Leja interpolation, all of which exhibit substantial growth as the number of grid points increases. It can also be seen that denoising results in a slightly higher number of overall iterations, so it is not beneficial for this problem. This is not surprising, as this is a diffusive problem that has a smooth solution and is therefore less susceptible to high-frequency oscillations during Lanczos iteration. Although AKP is slightly more accurate than KSS-EPI for fifth order, this is more than offset by the superior efficiency of KSS-EPI with denoising, particularly at larger grid sizes and larger time steps.

6.2. Advective problem

For our second test problem, we choose an advection-dominated PDE: a 1-D Burgers' equation

$$u_t + uu_x = \nu u_{xx}, \quad x \in [0,1], \quad t \in [0,1] \tag{32}$$

with $\nu = 0.03$. We impose homogeneous Dirichlet boundary conditions, and the initial condition

$$u_0(x) = \sin^3(3\pi x)(1-x)^{3/2}.$$

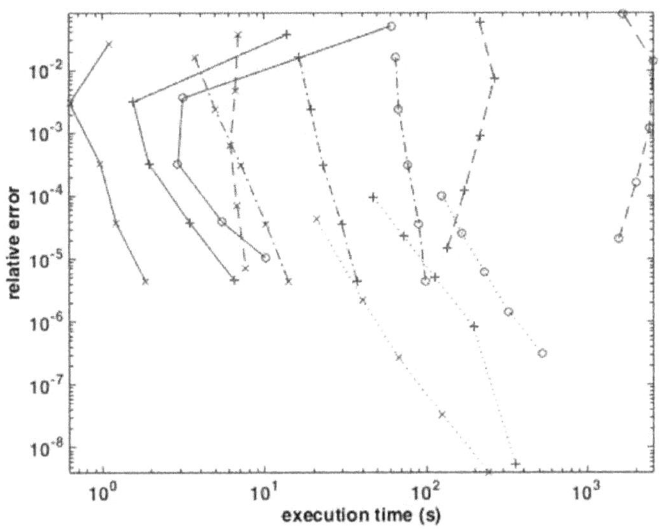

Figure 8. Relative error plotted against execution time for solving Burgers' equation (32) using the third-order EPI method (29). Matrix function-vector products are computed using KSS-EPI with denoising (solid curves), Krylov-EPI (dashed curves), AKP (dashed-dotted curves), and LEJA (dotted curves), on grids with N = 500 ('+' markers), 1500 ('x' markers), and 3000 ('o' markers) points. Time steps used are $\Delta t = (0.01)2^{-p}$, for p = 0, 1, 2, 3, 4.

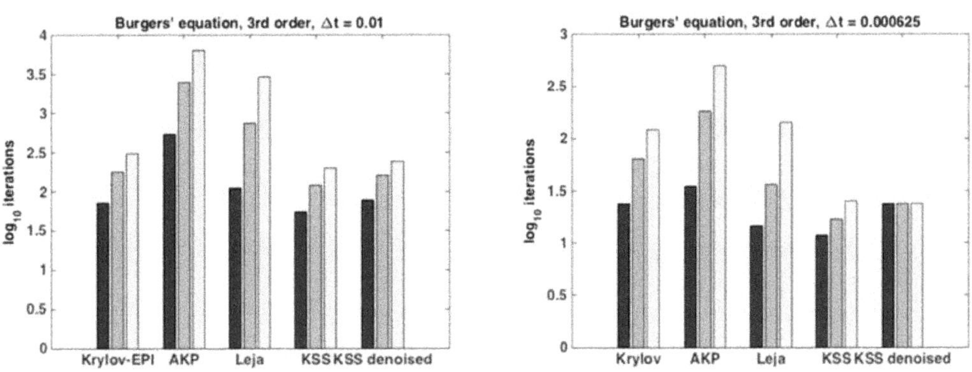

Figure 9. Average number of matrix-vector products, shown on a logarithmic scale, per matrix function-vector product evaluation for each method when solving Burgers' equation (32) using the third-order EPI method (29). For KSS and KSS denoised, FFTs are also included. For each method, bars correspond to grid sizes of N = 500, 1500, 3000 points, from left to right. Left plot: $\Delta t = 0.01$. Right plot: $\Delta t = 0.000625$.

For KSS-EPI, the low-frequency portion $\mathbf{b}_L$ consists of all components with wave numbers $\omega \leq N_c = 40$. The higher threshold for this problem, compared to the Allen-Cahn equation (31), is due to the fact that the initial data is less smooth.

The results are shown in Figures 8 and 9 for the third-order EPI method (29), and in Figures 10 and 11 for the fifth-order EPI method (30). From Figures 8 and 10, it can be seen that Krylov-EPI and KSS-EPI have comparable accuracy, but even for the coarsest grid with $N = 500$ points, KSS-EPI is much faster, and as with the Allen-Cahn equation, this gap in performance only grows with N. As shown in Figures 9 and 11, this is again due to the increasing number of Krylov projection steps needed for Krylov-EPI. While Leja interpolation is more accurate than KSS-EPI for both the third- and fifth-order EPI methods, and AKP is also more accurate in the fifth-order case, both of these methods are still significantly slower than KSS-EPI, and a similar scalability gap can also be observed.

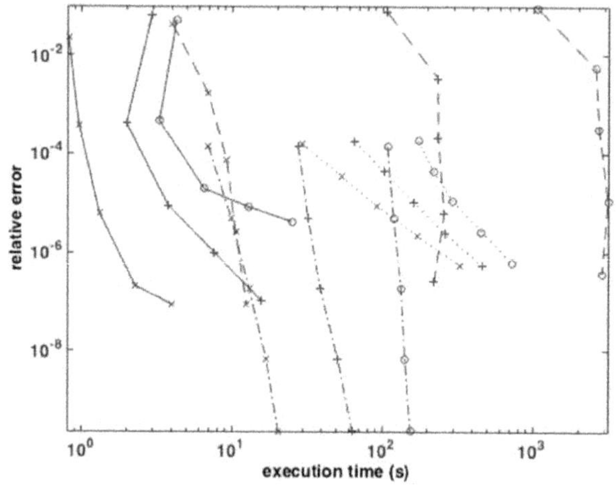

Figure 10. Relative error plotted against execution time for solving Burgers' equation (32) using the fifth-order EPI method (30). Matrix function-vector products are computed using KSS-EPI with denoising (solid curves), Krylov-EPI (dashed curves), AKP (dashed-dotted curves), and LEJA (dotted curves), on grids with N = 500 ('+' markers), 1500 ('x' markers), and 3000 ('o' markers) points. Time steps used are $\Delta t = (0.01)2^{-p}$, for p = 0, 1, 2, 3, 4.

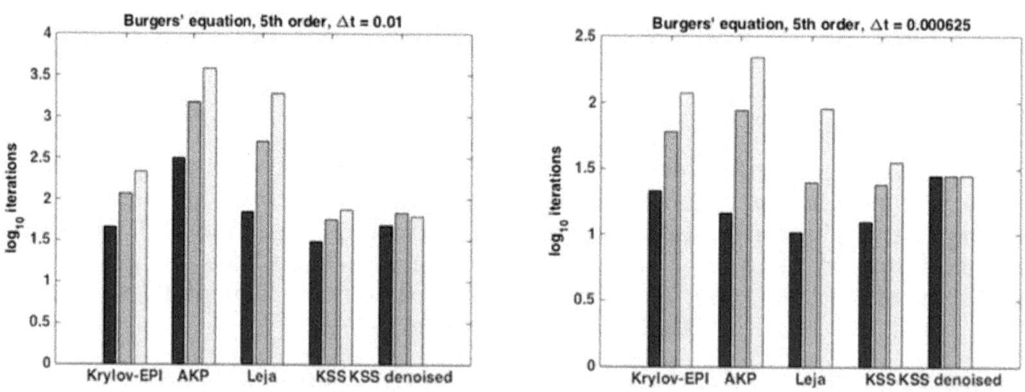

Figure 11. Average number of matrix-vector products, shown on a logarithmic scale, per matrix function-vector product evaluation for each method when solving Burgers' equation (32) using the fifth-order EPI method (30). For KSS and KSS denoised, FFTs are also included. For each method, bars correspond to grid sizes of N = 500, 1500, 3000 points, from left to right. Left plot: $\Delta t = 0.01$. Right plot: $\Delta t = 0.000625$.

The difference in scalability among the four approaches to computing matrix function-vector products is more apparent in Figures 9 and 11. Denoising applied to KSS-EPI is advantageous for this problem, unlike with the Allen-Cahn equation. As can be seen in these figures, for KSS-EPI without denoising, the number of iterations is increasing with the number of grid points, though not nearly as rapidly as with the other methods. However, with denoising included, the insensitivity of the number of iterations to the grid size is restored.

6.3. Discussion of efficiency

The major components of the computational cost of KSS-EPI stem from Krylov projection that is applied to low-frequency parts, and FFTs that are applied to both the low- and high-frequency parts. Specifically, suppose the EPI method is of order p, and q Krylov projection steps are needed for convergence of the low-frequency part. Then, the task of evaluating $\varphi(A\tau)b$ requires $q + p$ matrix-vector multiplications, $(p + 3)/2$ FFTs and 2 inverse FFTs if denoising is not used, and $q + p$ matrix-vector multiplications, $q + (p + 3)/2$ FFTs and $q + 2$ inverse FFTs if denoising is used. Denoising, therefore, is only worthwhile if the value of q can be substantially reduced, as in the case of Burgers' equation.

7. AT A GLANCE

We have demonstrated that when solving stiff systems of nonlinear ODEs derived from PDEs, the growth in the computational cost that results from an increase in the number of grid points can be significantly reduced by performing a relatively low and grid-insensitive number of Krylov projection steps and FFTs on low- and high-frequency portions of the solution separately, instead of a number of Krylov projection steps on the entire solution that grows substantially with the number of grid points. The component-wise approach employed by KSS methods, in which each Fourier coefficient of a matrix function-vector product $\varphi(\tau A)\mathbf{b}$ is computed using a frequency-dependent interpolant of φ, allows the Krylov subspace dimension to be bounded independently of the grid size and instead determined by the desired temporal order of accuracy (for the high-frequency part) and by the number of Krylov projection steps required on a coarse grid (for the low-frequency part).

Future work on KSS-EPI methods will focus on the computation of low-frequency Fourier components of the solution, as this step accounts for the most significant part of the running time. This work requires an efficient approach to automatically selecting the threshold N_c that is used to distinguish low-frequency from high-frequency components. Such adaptivity can be based on the smoothness of the solution and the performance of Krylov projection during previous time steps. In addition, computing low-frequency components with methods other than

Krylov projection, including Leja interpolation and adaptive Krylov projection, will be investigated. Such combination requires examination of error estimation and stopping criteria for these methods, to determine whether their convergence can be accelerated if it is known that the initial vector represents a bandlimited function.

It is important to note that the decomposition of the block Gaussian quadrature nodes into frequency-dependent and frequency-independent nodes is not limited to cases in which Fourier decomposition is used. In [26], the same idea is used for PDEs defined on a disk, in which a Legendre polynomial expansion is used for the radial part of the solution. The decoupling of the eigenvalue problem for T_K first observed in [11], and further exploited in [6, 7], occurs due to the rapid decay of the coefficients in whatever series expansion is used to represent the solution. Future work will include taking advantage of this behavior in order to generalize KSS methods to PDEs defined on non-rectangular domains.

8. CONCISE SUMMARY

The numerical solution of a time-dependent PDE generally involves the solution of a stiff system of ODEs arising from spatial discretization of the PDE. There are many methods in the literature for solving such systems, such as exponential propagation iterative (EPI) methods, that rely on Krylov projection to compute matrix function-vector products. Unfortunately, as spatial resolution increases, these products require an increasing number of Krylov projection steps, thus drastically increasing computational expense.

REFERENCES

1. Tokman, M.: Efficient integration of large stiff systems of ODEs with exponential propagation iterative (EPI) methods. J. Comp. Phys. 213 (2006) 748–776.
2. Loffeld, J., Tokman, M.: Comparative performance of exponential, implicit and explicit integrators for stiff systems of ODEs. Submitted.
3. Lambers, J. V.: Enhancement of Krylov subspace spectral methods by block Lanczos iteration. Electron. T. Numer. Ana. 31 (2008) 86–109.
4. Lambers, J. V.: An explicit, stable, high-order spectral method for the wave equation based on block Gaussian quadrature. IAENG Journal of Applied Mathematics 38 (2008) 333–348.
5. Guidotti, P., Kim, Y., Lambers, J. V.: Image restoration with a new class of forward-backward diffusion equations of Perona-Malik type. SIAM Journal on Imaging Sciences 6(3) (2013) 1416–1444.

6. Cibotarica, A.: Solution of Nonlinear Time-Dependent PDE Through Componentwise Approximation of Matrix Functions. Ph.D. Dissertation, The University of Southern Mississippi, 2015.
7. Cibotarica, A., Lambers, J. V., Palchak, E. M.: Solution of nonlinear time-dependent PDE through componentwise approximation of matrix functions. Submitted.
8. Hochbruck, M., Lubich, C.: A Gautschi-type method for oscillatory second-order differential equations. Numer. Math. 83 (1999) 403–426.
9. Hochbruck, M., Lubich, C.: On Krylov subspace approximations to the matrix exponential operator. SIAM J. Numer. Anal. 34 (1996) 1911–1925.
10. Hochbruck, M., Lubich, C., Selhofer, H.: Exponential integrators for large systems of differential equations. SIAM J. Sci. Comput. 19 (1998) 1552–1574.
11. Palchak, E. M., Cibotarica, A., Lambers, J. V.: Solution of time-dependent PDE through rapid estimation of block Gaussian quadrature nodes. Lin. Alg. Appl. 468 (2015) 233–259.
12. Golub, G. H., Meurant, G.: Matrices, moments and quadrature. Proceedings of the 15th Dundee Conference, June-July 1993, Griffiths, D. F., Watson, G. A. (eds.), Longman Scientific & Technical (1994)
13. Lambers, J. V.: Explicit high-order time-stepping based on componentwise application of asymptotic block Lanczos iteration, Numerical Linear Algebra with Applications 19(6) (2012) 970–991.
14. Atkinson, K.: An Introduction to Numerical Analysis, 2nd Ed. Wiley (1989)
15. Golub, G. H., Underwood, R.: The block Lanczos method for computing eigenvalues. Mathematical Software III, J. Rice Ed., (1977) 361–377.
16. Lambers, J. V. Krylov subspace spectral methods for the time-dependent Schrödinger equation with non-smooth potentials. Numerical Algorithms 51 (2009) 239–280.
17. Lambers, J. V.: Krylov subspace methods for variable-coefficient initial-boundary value problems. Ph.D. Thesis, Stanford University SC/CM program, 2003.
18. Lambers, J. V.: A spectral time-domain method for computational electrodynamics. Advances in Applied Mathematics and Mechanics 1(6) (2009) 781–798.
19. Saylor, P. E., Smolarski, D. C.: "Why Gaussian quadrature in the complex plane?" Numerical Algorithms 26 (2001) 251–280.
20. Golub, G. H., van Loan, C. F.: Matrix Computations, 3rd Ed., Johns Hopkins University Press (1996).

21. Saad, Y.: Numerical Methods for Large Eigenvalue Problems. Halsted Press, New York (1992).

22. Goodman, J., Hou, T., Tadmor, E.: On the stability of the unsmoothed Fourier method for hyperbolic equations. Numer. Math. 67(1) (1994) 93–129.

23. Bergamaschi, L., Caliari, M., Martnez, A., Vianello, M.: Comparing Leja and Krylov approximations of large scale matrix exponentials. Lecture Notes in Comput. Sci., 6th International Conference, Reading, UK, May 28–31, 2006, Proceedings, Part IV, Alexandrov, V. N., van Albada, G. D., Sloot, P. M. A., Dongarra, J. (eds.), Springer (2006) 685–692.

24. Rainwater, G., Tokman, M.: A new class of split exponential propagation iterative methods of Runge-Kutta type (sEPIRK) for semilinear systems of ODEs. J. Comp. Phys. 269 (2014) 40–60.

25. Tokman, M.: A new class of exponential propagation iterative methods of Runge-Kutta type (EPIRK). J. Comp. Phys. 230 (2011) 8762–8778.

26. Richardson, M., Lambers, J. V.: Krylov subspace spectral methods for polar and cylindrical geometries. In preparation.

02 ANALYTICAL METHODS FOR SOLVING GENERAL INVERSE EIGENVALUE PROBLEMS USING STRUCTURED APPROACHES

1. INTRODUCTION: AN OVERVIEW

Generalized inverse eigenvalue issues can be used to reframe many physical difficulties. In mechanical and structural system design, engineers frequently encounter the challenge of developing systems with natural frequencies that either need to be outside a specified range or precisely match certain frequencies. These design issues can be classified as inverse eigenvalue problems since the eigenvalues have a direct correlation with the natural frequencies of the system.

The study of discrete inverse eigenvalue issues has received significant attention. This has been made clear by a thorough review of the topic by Chu and Golub [1, 2]. Gladwell [3] takes a more direct route in which he considers specific inverse problems and matrix structures related to mechanical vibrations. Particularly, it appears that most of the literature focuses on system identification. One of the most common techniques in inverse eigenvalue problems is to use/measure the system's spectrum and then constrain the system in some fashion in order to obtain a second spectrum [4–8]. It is evident that there is typically a system in place that can be tested to gather the necessary data for the mathematical reconstruction of the system in question. While intriguing, this approach is not applicable for creating a new engineering design that aims to build a system with a specified spectrum without relying on another existing system as a basis for the design.

The predominant focus of research on inverse eigenvalue issues has been on the determination of a solution's existence, uniqueness, and computability. Additional research often involves modifications of the aforementioned experiments, including partially defined problems where not all spectrum information is known. These types of problems have been considered in the studies of Gladwell and Willms [9] and Ram and Elhay [10] and are of interest for problems requiring only certain frequencies to be specifically determined. Interestingly, Dias de Silva [11] and de Oliveira [12] have shown that an $n \times n$ matrix always exists when a minimum of $n - 1$ prescribed matrix entries and a prescribed characteristic polynomial are given as design information. The results of Dias de Silva and de Oliveira guarantee existence but not uniqueness of the matrix.

A major limitation of existing inverse eigenvalue theory is the absence of a solution for matrices with predetermined structures that cannot be solved using current recognized methods. This chapter presents two recently proposed strategies for directly solving inverse eigenvalue problems with structured solutions. The methods given are not limited to matrices of a specific kind and can therefore be used to matrices in general. The first method involves the development of the Cayley-Hamilton theorem for the generalized eigenvalue vibration issue. The Cayley–Hamilton theorem algorithm is demonstrated to be an effective design tool for resolving inverse generalized eigenvalue problems. Illustrations of the implementation of the method are provided. Another approach, known as the inverse eigenvalue determinant method, is also presented. This method offers an alternative and straightforward approach to reconstructing the matrices of the generalized eigenvalue problem. It requires prior knowledge of the eigenvalues and various physical parameters. Regarding the first method, there are no limitations on the permissible matrices for the inverse issue. Additionally, the technique is exemplified through several application scenarios, including examples that are specifically focused on practical implementation.

2. PROBLEM STATEMENTS

In this chapter, we consider the following problems:

- PROBLEM A: *Given a specified set of eigenvalues, $\lambda_1, \ldots, \lambda_n$ construct an n-order system, described by an $n \times n$ matrix A, which has $\lambda_1, \ldots, \lambda_n$ as its eigenvalues.*
- Although this problem has been considered for specific forms of matrices (i.e. Jacobi, band or other matrix forms), a general solution approach does not currently exist. Problem A leads to the generalized inverse eigenvalue problem, also presented here:
- PROBLEM B: *Given a specified set of eigenvalues, $\lambda_1, \ldots, \lambda_n$ construct a system with n-degrees of freedom, described by two $n \times n$ matrices (M and K), which has $\lambda_1, \ldots, \lambda_n$ as its generalized eigenvalues:* $\det(K - \lambda M) = 0$ *for the given* $\lambda_1, \ldots, \lambda_n$.
- For an engineer, Problem B relates directly to the design problem stated earlier, where a conservative vibrating system having specified natural frequencies is sought.

- PROBLEM C: *Given a specified set of eigenvalues,* $\lambda_1, \ldots, \lambda_m$ *construct a system with n degrees of freedom, where m < n, described by two n × n matrices (M and K), which has* $\lambda_1, \ldots, \lambda_m$ *as its generalized eigenvalues:* $\det(K - \lambda M) = 0$ *for the given* $\lambda_1, \ldots, \lambda_m$.
- Problem C, referred to as a partially described system, is one for which only a select few eigenvalues are specified, as opposed to the entire spectrum being given.

3. CAYLEY–HAMILTON APPROACH FOR INVERSE EIGENVALUE PROBLEMS

3.1. Basic Theory

In order to solve Problem A, a discussion of the Cayley–Hamilton theorem is required. The Cayley–Hamilton theorem states that if $p(\lambda)$ is the characteristic polynomial of a square matrix A, obtained from $p(\lambda) = \det(\lambda I - A)$, then substituting A for λ in the polynomial gives the zero matrix. Thus, by applying the theorem, matrix A satisfies its own characteristic polynomial, $p(A) = 0$ [13].

The Cayley–Hamilton theorem can be useful in inverse eigenvalue problems beyond the typical statement that a square matrix satisfies its own characteristic equation. Once the characteristic polynomial of a system is found from the specified spectral data, the Cayley–Hamilton theorem can be used to find an unknown matrix A, which represents the system. A set of unknown entries of the matrix A can be found from the set of equations that arise from the Cayley–Hamilton theorem. For example, suppose that a 2 × 2 matrix A is given by the form:

$$A = \begin{bmatrix} a_{11} & a_{12} \\ a_{21} & a_{22} \end{bmatrix} \qquad (1)$$

In order to solve the inverse eigenvalue problem, the entries of matrix A can be populated using the Cayley–Hamilton theorem in conjunction with the set of specified eigenvalues. In order to demonstrate the process, suppose that the specified eigenvalues are given by -2 and -3. The characteristic polynomial can then constructed as

$$p(\lambda) = (\lambda + 2)(\lambda + 3) = \lambda^2 + 5\lambda + 6 \qquad (2)$$

Using the Cayley–Hamilton theorem, λ in equation (2) is replaced by A from equation (1) so that

$$p(A) = A^2 + 5A + 6I = \mathbf{0} \qquad (3)$$

where **I** is the identity matrix and **0** is the zero matrix. Expressing equation (3) term-by-term gives four equations with four unknowns, specifically

$$\begin{bmatrix} a_{11}^2 + a_{12}a_{21} + 5a_{11} + 6 & a_{11}a_{12} + a_{12}a_{22} + 5a_{12} \\ a_{11}a_{21} + a_{21}a_{22} + 5a_{21} & a_{22}^2 + a_{12}a_{21} + 5a_{22} + 6 \end{bmatrix} = \begin{bmatrix} 0 & 0 \\ 0 & 0 \end{bmatrix} \quad (4)$$

However, equation independence is unclear. A discussion on this point is found in section 4.1. Solving the set of equations in equation (4) leads to two solutions:

$$\left[a_{11} = -a_{22} - 5,\ a_{12} = -\frac{a_{22}^2 + 5a_{22} + 6}{a_{21}} \right] \quad (5)$$

Any values can be assigned to a_{21} and a_{22} in equation (5), and the matrix A will have the desired eigenvalues given in equation (2). Consequently, Problem A has been solved, although it is clear that many solutions exist. This solution is particularly useful in solving inverse eigenvalue problems because it gives a range of A matrix values for which a system produces the same eigenvalues. The factors limiting the solutions as given in equation (5) are then based on the physical limits and fixed parameters of the system. Physical limits and fixed parameters serve to produce a set of solutions that are mathematically as well as physically possible.

3.2. Generalized Cayley–Hamilton theorem

Problem B is related to Problem A but takes on a more general form, which is conducive to real physical systems since many physical problems can be written as generalized eigenvalue problems. Although the characteristic polynomial for the generalized eigenvalue problem is similar to the single matrix case, it now involves two matrices rather than one. Therefore, the lesser known generalized Cayley–Hamilton theorem must be used by Chang and Chen [14].

The generalized Cayley–Hamilton theorem is modified to include a two square matrices, K and M. The nomenclature is used as a reminder that in many engineering applications the matrices represent mass (M) and stiffness (K) matrices. For the generalized eigenvalue problem, the characteristic polynomial takes the form $p(\lambda) = \det(K - \lambda M)$. By substituting K and M for λ into the characteristic polynomial, the following relationship must be satisfied.

$$p(K,M) = c_n\left(M^{-1}K\right)^n + c_{n-1}\left(M^{-1}K\right)^{n-1} \ldots + c_1\left(M^{-1}K\right) + c_0 I = 0 \quad (6)$$

where c_n is the coefficient of λ^n in $p(\lambda)$. Equation (6) is valid as long as M is non-singular.

If the matrices K and M commute (i.e. $KM = MK$), then the generalized Cayley–Hamilton theorem can be written as

$$p(K,M) = c_n K^n + c_{n-1} K^{n-1} M \ldots + c_1 K M^{n-1} + c_0 M^n = 0 \quad (7)$$

where no other restrictions are placed on matrix M. Solving these equations leads to the solution of Problem B.

3.3. Simple Numerical example

Once again, a simple numerical example is used to demonstrate concepts. Using the same desired eigenvalues as for the previous example gives the characteristic polynomial given in equation (2). This time, the modified generalized Cayley–Hamilton theorem is used to form an equation for unknown matrices *K* and *M* so that

$$p(K,M)=\left(M^{-1}K\right)^{2}+5\left(M^{-1}K\right)+6I=\mathbf{0} \qquad (8)$$

Assuming that the *K* and *M* matrices have a similar form to that of *A* in equation (1), then equation (8) can be expanded to give four equations (from the four matrix entries) with eight unknowns (four unknown entries in each of the two matrices). Solving these equations produces several results, one of which can be written as

$$\begin{bmatrix} k_{11}=\dfrac{k_{12}k_{21}k_{22}+5k_{12}k_{21}m_{22}+6k_{12}m_{21}m_{22}+6k_{21}m_{12}m_{22}-6k_{22}m_{12}m_{21}}{\left(k_{22}+2m_{22}\right)\left(k_{22}+3m_{22}\right)} \\ m_{11}=\dfrac{k_{21}k_{22}m_{12}+k_{12}k_{22}m_{21}-k_{12}k_{21}m_{22}+5k_{22}m_{12}m_{21}+6m_{12}m_{21}m_{22}}{\left(k_{22}+2m_{22}\right)\left(k_{22}+3m_{22}\right)} \end{bmatrix} \qquad (9)$$

Once again it becomes clear that only two solutions are dependent, and selecting any values for k_{12}, m_{12}, k_{21}, m_{21}, k_{22}, m_{22}, will result in *K* and *M* matrices that produce a solution with the desired eigenvalues.

Increasing the size of the square matrices increases the number of independent variables faster than it does the dependent variables. In other words, for an *n*th order system with *n*-specified eigenvalues, $2n^2$ unknown variables are required to find *K* and *M*, but, as will be shown later in this chapter, the generalized Cayley–Hamilton theorem only produces *n*-independent equations.

3.4. Algorithm

In order to use the Cayley–Hamilton theorem as a tool to find the *n* dimensional generalized eigenvalue system problem based on knowledge of desired eigenvalues and/or physical parameters, $2n^2$ - *n additional* pieces of design information are required in addition to the *n*-desired eigenvalues. An algorithm for solving Problem B is as follows:

ALGORITHM. Given the n-desired eigenvalues λ_1, ..., λ_n of an nth order generalized eigenvalue system

1. *Generate the characteristic polynomial:*
   $$p(\lambda)=(\lambda-\lambda_1)\ldots(\lambda-\lambda_{n-1})(\lambda-\lambda_n)=c_n\lambda^n+c_{n-1}\lambda^{n-1}+\ldots+c_1\lambda+c_0$$
2. *Generate forms for the M and K matrices from physical parameters and using the additional $2n^2$ - n pieces of design information, applying symmetry and any other techniques based on the discretized forward problem and leaving an n number of unknowns;*

3. *Generate the Cayley–Hamilton theorem equation:*

$$p(K,M) = c_n\left(M^{-1}K\right)^n + c_{n-1}\left(M^{-1}K\right)^{n-1}\ldots + c_1\left(M^{-1}K\right) + c_0 I = 0$$

or for commuting M and K matrices:

$$p(K,M) = c_n K^n + c_{n-1}K^{n-1}M\ldots + c_1 K M^{n-1} + c_0 M = 0$$

4. *Extract all n^2 equations from the Cayley–Hamilton equation in step 3;*
5. *Select n non-zero independent equations (the Cayley–Hamilton matrix diagonal entries work well);*
6. *Compute the n unknowns by solving the n-independent equations;*
7. *Insert the n computed values into their appropriate places in the M and K matrices;*
8. *Verify that a valid solution is obtained by calculating det(λM + K) = 0 and ensuring that the initially desired eigenvalues are obtained.*

The output consists of the n matrix entries of the M and K matrices.

4. ANALYSIS

In this section, results on the number of independent equations and required pieces of information used in the prior sections are discussed.

4.1. Information produced by the Cayley–Hamilton theorem

We show here that given n distinct eigenvalues for an nth order system, the Cayley–Hamilton theorem can produce at most n-independent equations, even though n^2 equations are produced through the entry-by-entry analysis. Let $p(t) = c_n t^n + c_{n-1} t^{n-1} + \ldots c_1 t + c_0$ be the characteristic polynomial of an $n \times n$ matrix A. The Cayley–Hamilton theorem states that $p(A) = c_n A^n + c_{n-1} A^{n-1} + \ldots c_1 A + c_0$ is the zero matrix, which gives n^2 equations. Now, suppose that the matrix A is diagonal and let the diagonal entries be $\lambda_1, \lambda_2, \ldots, \lambda_n$. The characteristic polynomial is

$$p(t) = (t-\lambda_1)(t-\lambda_2)\ldots(t-\lambda_n) \qquad (10)$$

In this case, since A is diagonal, $p(A)$ is also a diagonal matrix with exactly n equations. In fact, the ith diagonal entry is $p(\lambda_i)$.

Now, consider the case that A is not diagonal. Given n distinct eigenvalues for A, then A is diagonalizable, so that $A' = P^{-1}AP$ is diagonal for some invertible matrix P. The characteristic polynomial of A' is the same as the characteristic polynomial $p(t)$ of A. In fact, the diagonal matrix A' is the trivial solution to our design problem, and the 'structure' of the problem is housed in the matrix P. It is known that [15]

$$p(A) = P\; p(A')P^{-1} \qquad (11)$$

Equation (11) states that $p(A)$ can be obtained by combining the equations contained in $p(A')$. At the same time, since A' is itself diagonal, then matrix $p(A')$

contains exactly n equations. Thus, this states that $p(A)$ is a combination of exactly n-independent equations and we can expect that although $p(A)$ has n^2 equations, only n of them are independent.

It is also important to point out that the n-independent equations obtained from the Cayley–Hamilton theorem are each nth order polynomials. For a polynomial system with n unknowns and also n equations, then Bézout's theorem states that the problem has n^n complex solutions [16].

4.2. Required information

It is clear that the spectral information (eigenvalues) alone is not enough to solve the problem. The method presented, along with all other methods, is limited by the fact that an nth order system can produce at most n-independent equations, even though n^2 equations are obtained through applying the generalized Cayley–Hamilton theorem, as discussed above. If the matrices are completely unknown, then there may be as many as $2n^2$ unknown entries in the M and K matrices. The Cayley–Hamilton method can produce at most n-independent variables and the remaining equations must be specified in other ways. For creating physically realistic systems, this generally entails preconditioning or constraining the structure of the matrices.

For solving discrete conservative vibration problems, Equation (6) can be used as long as the M matrix is non-singular, and equation (7) can be used if M and K commute. It becomes very clear by considering the example in equation (9) that more information is required in order to build a suitable vibrating system based on M and K. From an engineer's point of view, any system, which can produce similar eigenvalues, has the potential of being a suitable solution, as long as it fits within the physical criteria set at the outset of the project. It is obvious that the example in Equation (9) can produce an infinite number of possible solutions. Equation (9) represents one of the several solutions to the Equation 8. As the system's order increases, so does the number of potential solutions (Bézout's theorem). Thus, engineers have many solutions at their disposal for creating suitable and optimized designs.

The inverse problem is then not limited by the eigenvalues, and in fact the eigenvalues alone do not contain enough information from which to build a physical system. Therefore, other information is required to solve the problem.

4.3. Structural constraints

One of the easiest ways to limit the number of matrix entries and include structural constraints is by including zero entries or by incorporating symmetry into the matrices. These constraints are interesting because they follow directly from real systems. For example, for many simply connected spring-mass systems, only diagonal entries are present in the mass matrix, and the stiffness matrix is expected to be symmetric and therefore can be mathematically forced to be symmetric by imposing that symmetric off-diagonal terms be the same. Physically, the structure of these matrices reflects the fact that the degrees of freedom of the system are coupled through stiffness element couplings only and no mass coupling is present. Furthermore, the symmetry of the stiffness matrix is a consequence of Newton's third law. The system is further constrained by the fact that the off-diagonal terms in the stiffness matrix are not independent, but are related to the diagonal terms, once again reducing the number of unknowns. In this case, incorporating physical constraints into the mathematical structure of the problem can reduce the number of unknowns from $2n^2$ to n^2.

In this case, the engineer still has the freedom to choose from several systems, which would satisfy the requirements. Further constraints may come in the form of available components, such as stiffeners, which must fit within specified physical dimensions or the financial budget.

These observations highlight the importance of the forward problem. Continuous systems are typically discretized using various methods, which produce a model of the system in matrix form. The form of the matrix is heavily dependent on the choice of discretization method. The inverse problem is consequently affected because it seeks to create a matrix that matches the form as chosen at the outset by the choice of discretization approach. Although the Cayley–Hamilton theorem does not discriminate in its ability to solve these various matrix forms, it is possible that certain discretization methods lead to simpler forms or matrices containing fewer variables. In this sense, an inverse problem may be easier to fully define.

4.4. Partially described systems

Another aspect that affects the amount of information required is the information available at the outset of the problem. Thus far, the entire spectral set has been used as input information, as well as specification of matrix entries when necessary. However, as stated in the study of Chu [1], often only portions of the entire spectrum are available. This is termed as a partially described inverse eigenvalue problem. This is true whether the subset of available spectral data arises from the design requirements or from the experimental data obtained for system identification. Regardless of the information available, the Cayley–Hamilton theorem can be used to produce n pieces of information for an nth order system stemming from n degrees of freedom. If certain eigenvalues are missing, the Cayley–Hamilton theorem can still be used.

Typically, in order to completely solve an nth order inverse problem using the Cayley–Hamilton theorem, only n unknown values should be present in the problem, regardless of the manner of their appearance along the solution path. Problem C makes full use of this detail during its solution by preconditioning the unknown matrix to account for this. For example, for a 3DOF system, it would be expected that only three unknowns should be present. Consider for example, the following system that has three known eigenvalues $\lambda_1 = 0.47, \lambda_2 = 4.66, \lambda_3 = 10.87$ and three unknown entries in the matrix given by

$$A = \begin{bmatrix} a_1 + a_2 & -a_2 & 0 \\ -a_2 & a_2 + a_3 & -a_3 \\ 0 & -a_3 & a_3 \end{bmatrix} \quad (12)$$

Using the Cayley–Hamilton theorem as described in Section 3, the matrix can be determined such that $a_1 = 9.43, a_2 = 1.23, a_3 = 2.06$, which gives a matrix solution for the system as

$$A = \begin{bmatrix} 10.66 & -1.23 & 0 \\ -1.23 & 3.28 & -2.06 \\ 0 & -2.06 & 2.06 \end{bmatrix} \quad (13)$$

As in Problem C, it is often the case that only partial spectral information is available. Therefore, the Cayley–Hamilton theorem can be used if the amount of missing spectral information is replaced by the same amount of matrix entry information. So, if only $\lambda_3 = 10.87$ is known then two of the three a matrix entries must be known. If $a_2 = 1.23$ and $a_3 = 2.06$ are known in the example above, then solving the Cayley–Hamilton equations gives $a_1 = 9.43, \lambda_1 = 0.47, \lambda_2 = 4.66$. Consequently, solving Problem C is not much different from solving Problem A. The same is true if the system is made up of a matrix M and a matrix K, except in this case, the generalized Cayley–Hamilton theorem must be used.

It becomes clear that one of the largest difficulties in using inverse eigenvalue theory to solve many physical problems is setting up the appropriate matrix form of the problem. This is achieved by consideration of the modelling approach chosen to model the forward problem. In most cases, the forward solution will utilize some form of discretization, such as finite elements, global elements, finite difference, and so on. The method chosen has a large impact on the structure of the matrix, making the solution easier or more difficult depending on the situation. Also, in discretizing, the method chosen to relate material parameters has a large effect on the number of independent variables. Therefore, it is extremely important to ensure that simplification methods are properly considered.

4.5. Polynomial solvers

An nth order system is represented by a characteristic equation that is a polynomial of nth order. The Cayley-Hamilton theorem will yield n linearly independent equations, as previously mentioned. In order to find the unknown quantities, a system of nth order polynomials must be solved for both partially described and completely described systems. To solve this, a proficient polynomial solver or polynomial root-finding technique will be necessary. The problems in the rest of this chapter were solved using the symbolic computer algebra software Maple, developed by Maplesoft in Waterloo, Canada. However, any modern numerical or symbolic software package might also be used to answer these problems. In our simulations, we utilized the Maple built-in function solve, together with the freely accessible package Direct Search. The specifics of the Direct Search package can be located in the research conducted by Moiseev [17]. Direct Search yielded a significantly higher number of valid solutions compared to solve. It is important to note that most of these issues do not have a unique solution, hence several solutions are anticipated. The presence of various solutions is advantageous in engineering design as it indicates the availability of numerous viable solutions. It is then the responsibility of the designer to select the most feasible solution that adheres to the given problem limitations. In order for the Cayley-Hamilton theorem (and the determinant approach discussed in the remaining part of the chapter) to be efficient, it is necessary to have a reliable polynomial solver that can accurately find all the potential solutions.

5. EXAMPLE WITH THE CAYLEY–HAMILTON INVERSE METHOD

In this section, we demonstrate by way of example how the Cayley–Hamilton method can be used to solve a practical engineering design problem by writing it as an inverse vibrations problem and thus as an inverse eigenvalue problem.

5.1. Problem statement

Given a desired fundamental frequency, construct a brace–plate system as described by a mass matrix M and a stiffness matrix K. All dimensional (geometric) properties of the brace–plate system are assumed to be specified and fixed except for the thickness of the brace h_c and the design variable, which must be found. The plate and brace material properties are also assumed to be known.

It is known from vibrations theory that the natural frequencies of a system specified by stiffness matrix K and mass matrix M are the square roots of the generalized eigenvalues of the matrices (K, M). Hence, a specified fundamental frequency of the system is the same as specifying the smallest generalized eigenvalue of (K, M).

5.2. Forward model

The first step in the solution process is to generate a finite-dimensional model of the continuous problem. The model is based on an orthotropic plate structurally reinforced by a brace in the weaker plate direction. The model is shown in Figure 1.

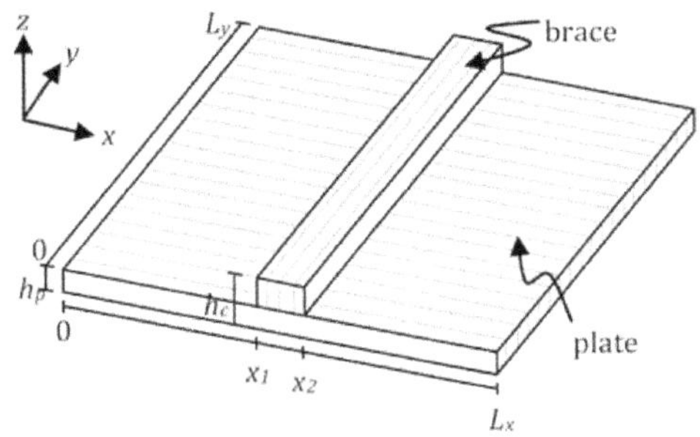

Figure 1. Orthotropic plate reinforced with a rectangular brace.

The forward model is discretized using the assumed shape method. The assumed shape method is an energy method, which uses global plate elements with the kinetic and strain energy plate equations in order to determine the system's equations of motion, from which the mass and stiffness matrices are extracted [18]. For the details of the development of the large mass and stiffness matrices, the reader is referred to the study of Dumond and Baddour[19]. The system is assumed simply supported, conservative and the material properties are assumed orthotropic. The forward model is created assuming that the mechanical properties are all related to Young's moduli in the y direction.

5.3. Inverse model

The goal is to reconstruct the brace-plate system from a desired fundamental frequency. The generalized Cayley-Hamilton theorem inverse eigenvalue method is used as explained in Section 3.2.

A cross section of the fundamental modeshape is shown in Figure 2. It is clear that the brace affects the maximum amplitude of this modeshape, thus also affecting the associated frequency. In order to adjust the fundamental frequency of the brace-plate system to a desired value, it is necessary to adjust the thickness of the brace [20].

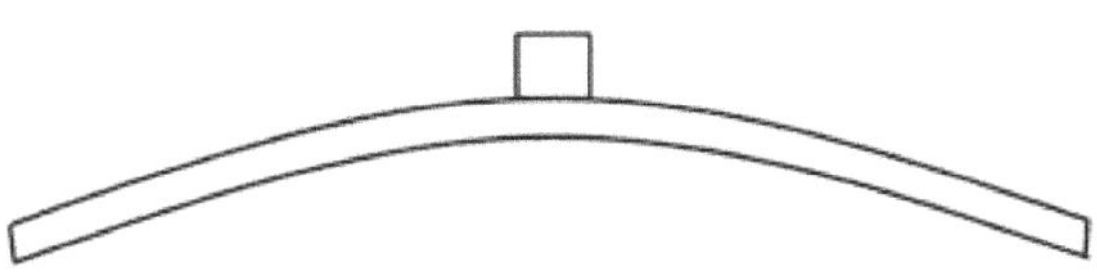

Figure 2. Cross section of the brace-plate system's fundamental modeshape.

5.4. Modelling considerations

The mechanical properties vary based on E_y, the Young's modulus in the y direction, and the brace thickness controls the brace-plate system's fundamental frequency. Therefore, the forward model is created using the assumed shape method while leaving these two parameters as variables. Thus, the mass matrix M is a function of h_c, the height of the brace-plate system at the (assumed fixed) location of the brace, and the stiffness matrix K is a function of h_c and also E_y. Here, 2×2 trial functions are used in the assumed shape method, so it follows that fourth order square matrices are created. The trial functions used are those of the simply supported rectangular plate given by

$$w(x,y,t)=\sum_{n_x=1}^{m_x}\sum_{n_y=2}^{m_y}\sin\left(\frac{n_x\pi x}{L_x}\right)\sin\left(\frac{n_y\pi y}{L_y}\right)q_{n_xn_y}(t) \tag{14}$$

where m are the modal numbers, q is the time function and w is the displacement variable normal to the plate. The displacement variable w is then used directly in creating the kinetic and strain energy equations of the simply supported rectangular plate. These equations are broken into three sections as shown in Figure 1 in order to take into account presence of the brace. This procedure is described in detail in the study of [19]. It is assumed that the E_y is known and used as input information into the stiffness matrix. This leaves h_c as the only unknown parameter, appearing in both the mass and stiffness matrices.

In order to solve these matrices from the desired fundamental frequency, we must first create the characteristic polynomial using the desired frequency,

$$p(\lambda)=(\lambda-a)\cdot(\lambda-b_1)\cdot(\lambda-b_2)\cdot(\lambda-b_3) \tag{15}$$

where a is the specified fundamental frequency, and b_1, b_2 and b_3 are unknown values, which will also need to be found as part of solving the problem. Since we have assumed 2×2 trial functions, so that the mass and stiffness matrices are both 4×4, the characteristic polynomial must be fourth order, as shown in equation (15). Subsequently, $p(\lambda)$ is expanded so that the polynomials coefficients can be found. Once the polynomial is created, the Cayley–Hamilton equation can be written by substituting $(M^{-1}K)$ for λ into equation (15):

$$p(K,M)=c_4\left(M^{-1}K\right)^4+c_3\left(M^{-1}K\right)^3+c_3\left(M^{-1}K\right)^2+c_1\left(M^{-1}K\right)+c_0I=0 \tag{16}$$

where c_n are the coefficients of λ in $p(\lambda)$ determined through equation (15). Equation (16) produces 16 equations, of which only 4 are independent. According to Bézouts theorem, selecting and solving the equations on the main diagonal for the four unknowns (h_b, b_1, b_2 and b_3) produce 4^4=256 possible solutions. From the set of all possible solutions, complex-valued solutions can be immediately eliminated as not being physically meaningful. Clearly, further constraints must be added to the solution in order to get a solution which fits within the desired physical limits. These physical limits are based on the maximum and minimum brace dimensions, which are required to compensate for the range of plate stiffnesses used during the analysis, as well as the range of natural frequencies, which can be obtained using these system dimensions. Thus, the following constraints are implemented into the solution:

$$
\begin{aligned}
&0.013 \le h_b \le 0.016 \text{ m} \\
&1 \times 10^7 \le b_1 \le 9 \times 10^8 \text{ rad / s} \\
&1 \times 10^7 \le b_2 \le 9 \times 10^8 \text{ rad / s} \\
&1 \times 10^7 \le b_3 \le 9 \times 10^8 \text{ rad / s}
\end{aligned} \tag{17}
$$

Solving the four equations obtained from equation (16) within the constraints provided by equation (17) yields, a physically realistic solution satisfies the desired fundamental frequency, as well as the system's parameters.

5.5. Results

The material properties of the brace and plate used during the analysis are given in Table 1.

Table 1. Material properties.

Material properties	Values
Density, μ (kg/m³)	403.2
Young's modulus, E_y (MPa)	850
Young's modulus, E_x (MPa)	$E_y/0.078$
Shear modulus, G_{xy} (MPa)	$E_x \times 0.064$
Poisson's ratio, ν_{xy}	0.372
Poisson's ratio, ν_{yx}	$\nu_{LR} \times E_y/E_x$

The dimensions used for the model throughout the analysis of the brace–plate system are shown in Table2.

Table 2. Dimensions of brace-plate model.

Dimensions	Values
Length, L_x (m)	0.24
Length, L_y (m)	0.18
Length, L_b (m)	0.012
Reference, x_1 (m)	$L_x/2-L_b/2$
Reference, x_2 (m)	x_1+L_b
Thickness, h_p (m)	0.003
Thickness, h_b (m)	0.012
Thickness, h_c (m)	h_p+h_b

These dimensions refer to those shown in Figure 1, where '*p*' refers to the plate's dimensions, '*b*' refers to the brace's dimensions and '*c*' refers to the dimensions of the combined system.

As a basis for comparison, a plate having a Young's moduli of E_y=850 MPa to which a brace is attached with a combined brace–plate thickness of h_c=0.015 m is found to have a fundamental natural frequency of 687 Hz, as calculated using the forward model. The analysis is then performed using the inverse Cayley–Hamilton method. As E_y of the plate is varied, the thickness of the brace-plate section is calculated such that the fundamental frequency of the brace–plate system is kept consistent at 687 Hz. The results of the computations are presented in Table 3.

Table 3. Results of the inverse model analysis.

Young's modulus E_R (MPa)	Brace thickness h_c (m)	Fundamental frequency a (Hz)
750	0.01576	687
800	0.01536	687
813	0.01527	687
850	0.01500	687
900	0.01466	687
950	0.01435	687

Clearly, adjusting the thickness of the brace also has an effect on the other natural frequencies. These are presented in Table 4.

Table 4. Calculated frequencies of the inverse model analysis.

Young's modulus E_R (MPa)	Brace thickness h_c (m)	b_1 (Hz)	b_2 (Hz)	b_3 (Hz)
750	0.01576	774	1360	2653
800	0.01536	782	1363	2650
813	0.01527	784	1364	2650
850	0.01500	790	1366	2648
900	0.01466	798	1370	2645
950	0.01435	806	1374	2642

Interestingly, the constraints indicated in equation (17), although physically strict, allow for more than one solution in certain cases. An example is shown in Table 5.

Table 5. Alternate brace thickness solution satisfying the physical constraints.

Young's modulus E_R (MPa)	Brace thickness h_c (m)	a (Hz)	b_1 (Hz)	b_2 (Hz)	b_3 (Hz)
750	0.01359	687	570	1149	2185

5.6. Discussion

From these results, it is evident that it is possible to design a brace–plate system starting with a desired fundamental frequency in conjunction with using the proposed Cayley–Hamilton method. Table 3 clearly shows that by adjusting the thickness of the brace by small increments (10^{-5} m, machine limit), it is possible to compensate for the variation in the cross-fibre stiffness (E_y) of the plate so that the fundamental frequency of the combined system is equal to that of the benchmark value of 687 Hz. The results obtained using the Cayley–Hamilton theorem algorithm match those values obtained using the forward model exactly. However, since no account has been taken of the other frequencies during the analysis, Table 4 shows that frequencies b_1–b_3 vary considerably from those values obtained for E_y=850 MPa. Therefore, it is important to ensure that there is a good understanding of what a given model can control. Moreover, it is interesting to note that within the strict physical constraints of equation (17), there is more than one brace-plate system (solution) that satisfies the Cayley–Hamilton theorem of equation (16). From Table 5, it can be seen that an alternate solution to the system exists, different from the one presented in Table 3, for a plate having an E_y of 750 MPa. In this case, by reducing the thickness of the brace, it is still possible to achieve a system having the desired frequency of 687Hz. However, the desired frequency is no longer the fundamental frequency but rather becomes the second frequency, and the fundamental has been replaced with a fundamental frequency of 570 Hz. It is important to keep this phenomenon in mind while designing a system. This is especially true if the order in the spectrum of a certain frequency associated with a certain modeshape is absolutely critical.

6. THE DETERMINANT METHOD FOR PARTIALLY DESCRIBED SYSTEMS

Although effective, the Cayley–Hamilton method for solving inverse eigenvalue problems described in the previous section may not be the most computationally efficient method when solving partially described systems. Partially described systems are those for which only a select few eigenvalues are specified, as opposed to the entire spectrum being given. In this section, the determinant method, which first appeared in the study of Mir Hosseini and Baddour [21], is introduced.

For clarity, the method is explained for the case where two eigenvalues are required to be a certain value. The method proceeds in the same manner where the number of specified eigenvalues is larger. Solving the characteristic equation of the system requires the use of the determinant equation

$$p(t) = \det(K - \lambda M) = 0 \qquad (18)$$

where λ is the eigenvalue obtained by finding the roots of the polynomial equation 18. Suppose that two of the possible n roots (in this case the eigenvalues a_1 and a_2) are required to be of a certain value. Replacing λ in the characteristic polynomial with the desired a_1 and a_2 each in turn leads to two scalar equations, specifically

$$\begin{aligned} p(a_1) &= \det(K - a_1 M) = 0 \\ p(a_2) &= \det(K - a_2 M) = 0 \end{aligned} \qquad (19)$$

The two scalar equations in equation (19) can be simultaneously solved for up to two unknowns. This greatly reduces the computational effort involved. Constraints can be used to create a physically realistic system. Similar to the case for the Cayley–Hamilton theorem, the equations will be polynomials and the use of a good polynomial solver is required to obtain the multiple solutions expected.

6.1. Algorithm

In the case of Problem C, m eigenvalues are given for an n dimensional system, so that $n - m$ is the number of unknown or unspecified eigenvalues for the system. The output of the determinant method consists of the remaining $n - m$ eigenvalues as well as any of the unknown matrix entries.

Determinant Algorithm. *A certain set of desired eigenvalues, $a_1, ..., a_m$, which is smaller than the dimension of the system, are given as input information.*

- *Generate the M and K matrices from the chosen forward modelling technique leaving variable m as unknown modelling parameters.*
- *Generate a determinant equation from each desired frequency:*

$p(a_1) = det(K - a_1M)$ and $p(a_2) = det(K - a_2M)...p(a_m) = det(K - a_mM)$where $a_1, ..., a_m$ are obtained from the desired natural frequencies;

- *Solve the set of m equations in m unknowns;*
- *Insert the unknown values found from step 3 into the matrices and calculate the system natural frequencies using the forward modelling process in order to verify that the solution is consistent with the input values.*

7. EXAMPLE OF USING THE DETERMINANT METHOD

In this section, an example of how the determinant method can be used to solve an inverse vibration problem is presented. In this section, the authors demonstrate a method to impose two desired natural frequencies on a dynamical system of a simply supported Euler–Bernoulli beam to which a single lumped mass is attached by determining the magnitude and mounting position of the mass. The full details of this example are given in the study of Mir Hosseini and Baddour [21].

7.1. Modelling

7.1.1. The method of assumed modes

The method of assumed modes was used to find a finite-dimensional model (discrete model) for the beam with attached elements. As is well known for the assumed modes approach, the greater the number of modes used, the more accurate the results obtained. Usually, the vibrational modes of a related but simpler problem are used to find approximate solutions to a more complicated problem, in this example, the modeshapes of the beam with no attached elements are used. A good introduction to the assumed modes can be found in the study of Meirovitch [22].

The assumed method, when applied to a continuous, conservative vibrational system, will result in mass and stiffness matrices that represent the system. The dimensions of these matrices are determined by the degree of discretization selected for the problem. Here lies the main advantage of the assumed mode method over the popular finite element analysis (FEA). It has been shown in the study of Cha [23] that the same level of accuracy can be reached by the assumed modes method using smaller degrees of discretization than with FEA. This implies mass and stiffness matrices that are smaller. This reduced dimensionality of the system is far more relevant to solution of the inverse problem than to the forward problem.

The assumed modes method was chosen to derive the equations of motion for the case of an Euler–Bernoulli beam to which a number of discrete elements are attached. The transverse vibrations of the beam can be written as

$$w(x,t)=\sum_{j=1}^{N}\phi_j(x)\,\eta_j(t) \tag{20}$$

Here, $w(x, t)$ is the transverse displacement of the beam, $\phi_j(x)$ is a space-dependent eigenfunction, $\eta_j(t)$ is the generalized coordinate and N is the number of assumed modes chosen for the problem. It is important to note that $\phi_j(x)$ varies with the choice of the beam and any $\phi_j(x)$ should be chosen to satisfy the required geometric boundary conditions of the selected beam.

7.1.2. Derivation of equations of motion

In order to derive the equations of motion for the one-dimensional Euler–Bernoulli beam with multiple lumped point-mass attachments as shown in Figure 3, expressions for kinetic and potential energies must first be found. The kinetic energy of the beam is given by

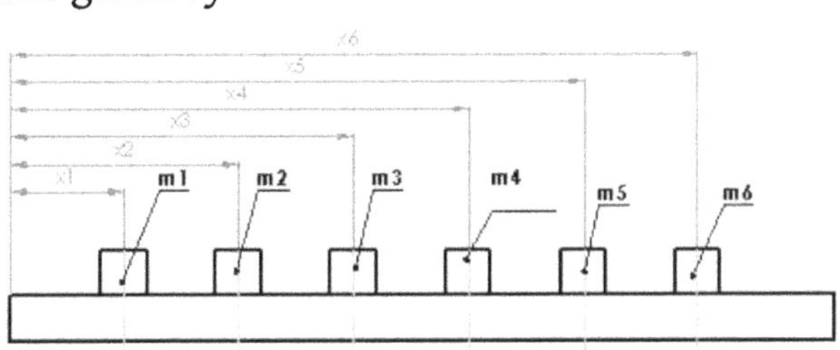

Figure 3. Beam with multiple lumped mass attachments.

$$T=\frac{1}{2}\sum_{j=1}^{N}M_j\dot{\eta}_j^2(t)+\frac{1}{2}m_1\dot{w}^2(x_1,t)+\ldots+\frac{1}{2}m_s\dot{w}^2(x_s,t) \tag{21}$$

where M_j is generalized masses of the bare beam (no attachments), an over dot indicates derivatives with respect to time and $m_1...m_s$ are s lumped point masses positioned at $x_1...x_s$, respectively.

Using the same procedure, the equation for the potential energy can be written as

$$V = \frac{1}{2}\sum_{j=1}^{N} K_j\ \eta_j^2(t) \qquad (252)$$

where K_j are the generalized stiffnesses of the bare beam.

Substituting equation (20) into equation (21), the following equation for kinetic energy is obtained

$$T = \frac{1}{2}\sum_{j=1}^{N} M_j\dot{\eta}_j^2(t) + \frac{1}{2}m_1\left[\sum_{j=1}^{N}\phi_j(x_1)\dot{\eta}_j(t)\right]^2 + \ldots + \frac{1}{2}m_s\left[\sum_{j=1}^{N}\phi_j(x_s)\dot{\eta}_j(t)\right]^2 \qquad (23)$$

Equation (22) for the potential energy remains unchanged from a bare beam since no elastic element was added to the beam.

Having found the expressions for kinetic and potential energies in terms of ϕ and η , these are then substituted into the Lagrange's equations to yield the equations of motion. Lagrange's equations are given by

$$\frac{d}{dt}\left(\frac{\partial T}{\partial \dot{\eta}_i}\right) - \frac{\partial T}{\partial \eta_i} + \frac{\partial V}{\partial \eta_i} = 0 \qquad i = 1, 2, .., N \qquad (24)$$

where N corresponds to the number of generalized coordinates and hence the number of differential equations.

Substituting equations (23) and (22) into equation (24) and converting the system of equations into a matrix representation, the matrix equation of motion is given by

$$\mathbf{M}\underline{\ddot{\eta}} + \mathbf{K}\underline{\eta} = \underline{0} \qquad (25)$$

where **M** and **K** are the system mass and stiffness matrices, respectively. The mass matrix is given by

$$\mathbf{M} = \mathbf{M}^d + m_1 \cdot \underline{\phi_1} \cdot \underline{\phi_1}^T + \ldots + m_s \cdot \underline{\phi_s} \cdot \underline{\phi_s}^T \qquad (26)$$

In equation (26), $\phi_1 \ldots \phi_s$ are N-dimensional column vectors of the N eigenfunctions evaluated at the attachment points $x_1 ... x_s$, so that for example

$$\underline{\phi_1} = \begin{bmatrix} \phi_1(x_1) \\ . \\ . \\ . \\ \phi_N(x_1) \end{bmatrix} \qquad (27)$$

$\mathbf{M}^d$ is a diagonal matrix whose diagonal components are the generalized masses M_i, and $m_1 ... m_s$ are the masses of the lumped attachments.

For the stiffness matrix, since elastic elements are not added to the beam, the stiffness matrix remains a diagonal matrix whose elements are the generalized stiffnesses of the beam. Hence, the stiffness matrix is given by

$$\mathrm{K}=\mathrm{K}^{\mathrm{d}} \qquad (27)$$

7.1.3. Choice of eigenfunctions

As can be seen in equations (26) and (27), eigenfunctions constitute a major component of the mass matrix of the combined system. It is solely a function of position, *x* and can take many different forms depending on the beam being considered. The most basic requirement of any eigenfunction simulating the vibrational behaviour of a beam is its ability to accommodate and satisfy the beam boundary conditions. Here, the eigenfunctions for an Euler–Bernoulli beam with simply supported boundary conditions are given by Cha [23]

$$\phi_n(x) = \sqrt{\frac{2}{\rho L}} \sin\left(\frac{n\pi x}{L}\right) \qquad (27)$$

where ρ is the mass per unit length of the beam and *L* represents the length of the beam.

7.1.4. Frequencies and modeshapes

In order to solve equation (25), a system of *N* second-order differential equations, the vector of generalized coordinates $\underline{\eta}$ is written as

$$\underline{\eta} = \vec{\eta} e^{i\omega t} \qquad (27)$$

Here, ω is the frequency of vibration of the system. Also, the inclusion of the complex number "*i*" is justified given the fact that the system is conservative, and it is expected that the vibrations are purely oscillatory.

Substituting equation (30) into equation (25) and taking derivatives yields

$$\left(-\omega^2\mathbf{M} + \mathbf{K}\right)\vec{\eta} = \underline{0} \qquad (28)$$

In order for equation (31) to have a non-trivial solution, the following equation must hold

$$\det\left(-\omega^2\mathbf{M} + \mathbf{K}\right) = 0 \qquad (29)$$

Substituting $\lambda = \omega^2$ in equation (32), an equivalent expression is obtained:

$$\det\left(\mathbf{K} - \lambda\mathbf{M}\right) = 0 \qquad (30)$$

Equation (33) can be solved for either the squared frequencies of the system λ (forward problem) or the coefficients of **K** and **M** matrices (inverse problem). In this section, the case of inverse problem is investigated for the case of an Euler–Bernoulli beam with a single mass attachment det(**K**).

7.1.5. Inverse problem

The problem of imposing two frequencies on a dynamical system consisting of a beam with a single attached lumped mass is considered here which leads to the following system of equations:

$$\begin{cases} \det(\mathbf{K} - \lambda_a \mathbf{M}) = 0 \\ \det(\mathbf{K} - \lambda_b \mathbf{M}) = 0 \end{cases} \quad (31)$$

where λ_a and λ_b are the desired natural frequencies squared to be imposed on the system (design variables). Due to the fact that no stiffness element is added to the beam, **K** remains intact and can be determined using equation (28), whereas the mass matrix of the combined system **M** is given by

$$\mathbf{M} = \mathbf{M}^d + m_1 . \underline{\phi_1} . \underline{\phi_1}^T \quad (32)$$

where m_1 is the magnitude of the lumped mass and ϕ_1 is a function of the mass position x_1 and is defined by equation (27). m_1 and x_1 are the unknown variables of the inverse problem. Substituting equation (35) into (34), a system of two equations and two unknowns is obtained whose solution is presented in the next section for the case of a simply supported beam.

In using the determinant method, each desired natural frequency is substituted into a determinant equation of the form of equation (34). Therefore, each desired natural frequency produces a single equation. Therefore, n desired natural frequencies require n "design degrees of freedom" or parameters to be controlled, such as added masses or springs, or their location on the beam. The value of the added mass (or spring) is considered one design degree of freedom and its unknown location is another. The form of the equations produced through imposing a desired frequency on the determinant depends on the choice of eigenfunctions in the assumed modes method and also how the unknown desired parameter (e.g. position of the mass) is included in those eigenfunctions. Here, we have chosen the traditional trigonometric functions as the eigenfunctions because they are the most well-known choices for the Euler–Bernoulli beam. Other choices of eigenfunctions may lead to better computational results with this method.

7.2. Results

The following assumptions are considered in defining the inverse problem:

- The acceptable mass range is a fraction of the mass of the beam, that is, $m = c\rho L$ for $0 < c < 1$.

- The degree of discretization using assumed modes is *N=10* for the simply supported beam.
- The acceptable position range is a fraction of the length of the beam *L*, that is, *pL*, where $0 \le p \le 1$.

Maple (Maplesoft) was used to perform the numerical calculations. The major steps in solving, as well as coding, the inverse eigenvalue problems are outlined here. First, two natural frequencies are chosen as desired input frequencies (λ_a, λ_b). These are the frequencies we seek to impose on the beam with its mass attachment. To insure a solvable problem, we chose known values from previously solved forward problem. The generalized mass and stiffness matrices must be formed. ($\mathbf{M}^d$, $\mathbf{K}^d$) whose diagonal elements are given by

$$M_i = 1$$

$$K_i = \frac{(i\pi)^4 EI}{(\rho L^4)} \qquad (33)$$

The eigenfunction vector, $\underline{\phi}_i$, must also be built using the simply supported beam eigenfunction equation (29). The stiffness matrix is unaffected because no stiffness elements are added. The mass matrix of the combined system is affected by the presence of lumped masses, whereas the stiffness matrix is not. Hence, the matrices are given by

$$\mathbf{K}_t = \mathbf{K}^d$$

$$\mathbf{M}_t = \mathbf{M}^d + c\rho L \underline{\phi}_{pL} \underline{\phi}_{pL}^T \qquad (34)$$

where *c* is the mass coefficient and vector $\underline{\phi}_{pL}$ pL is defined as

$$\underline{\phi}_{pL} = \begin{bmatrix} \phi_1(pL) \\ \cdot \\ \cdot \\ \phi_N(pL) \end{bmatrix} \qquad (35)$$

Equations (38) and (39) along with the two desired values of λ are substituted into equation (34). This gives two equations in two unknowns, which shall be solved for the two unknowns, *c* and *p*, using *fsolve* as well as the *DirectSearch* packages in Maple. *Fsolve* is Maple's built-in equation-solving package. The details of the *DirectSearch* package can be found in the study of Moiseev [17]. The results obtained for *c* and *p* include the anticipated results (already known from the forward problem since we chose the desired λ from a known forward problem) plus additional results for

c and p. To check whether the order of the frequencies in the frequency spectrum will be conserved or if the results returned by the inverse problem achieve the desired system frequencies, the parameters c and p must be substituted into the forward code in order to obtain the entire spectrum of system frequencies. From symmetry of the boundary condition about the beam midpoint, only half of the beam is considered and the results can be extended to the other half. These results are presented in the left-hand columns of Table 7.

Table 6. Inverse problem for second and third frequencies for a simply supported beam.

Input given to inverse determinant method	Solution via inverse determinant method	Full span of frequency squared spectrum obtained through the solution of forward problem
λ_2 = 1210.7430, λ_3 = 7703.7093 Obtained with $m = 0.2\rho L$ $l = 0.3L$ in the forward code)	$m = 0.2\rho L$ $l = 0.3L$ (This result was obtained through both Direct Search and solve methods)	76.8682 1210.7430 7703.7093 22705.2481 49733.6143 1.19385×10^5 2.29571×10^5 3.45310×10^5 5.95631×10^5 9.74091×10^5
	$m = 0.3895\rho L$ $l = 0.3587L$ (This result was obtained through both Direct Search and fsolve packages)	59.0673 1210.7430 7703.7093 18122.3022 55694.1292 1.18846×10^5 1.91164×10^5 3.88304×10^5 5.89699×10^5 8.69163×10^5

Table 7. Inverse problem for second and fourth frequencies for a simply supported beam.

Input given to inverse determinant method	Solution via inverse determinant Method	Full span of frequency squared spectrum obtained via the solution of forward problem
λ_2 = 1411.9003, λ_4 = 20420.5236 (Obtained with m = 0.2ρL l = 0.4L in the forward code)	m = 0.1516ρL l = 0.0972L (This result was obtained through **Direct Search** and **fsolve** packages)	94.7851 **1411.9003** 6626.2231 **20420.5236** 51205.1150 1.11210×10^5 2.15698×10^5 3.82314×10^5 6.29638×10^5 9.73288×10^5
	m = 0.4695ρL l = 0.4268L (This result was obtained through **Direct Search** package only)	51.2467 **1411.9003** 6082.1204 **20420.5235** 57662.4808 1.00203×10^5 2.33797×10^5 3.33408×10^5 6.17507×10^5 9.04206×10^5
	m = 0.2ρL l = 0.4L (This result was obtained via **Direct Search** package only)	71.3527 **1411.9003** 7164.6511 **20420.5236** 60880.6819 1.05946×10^5 2.21752×10^5 3.77388×10^5 5.73630×10^5 9.74091×10^5

The squared frequencies in the left-hand columns of these tables are chosen from the previously solved forward problem, and their subscripts indicate their order in the frequency spectrum. The middle columns of these tables contain the values of the masses as well as their positions on the beam obtained after substituting the squared frequencies of the first columns into the inverse problem code. Finally, the right-hand- columns are the full span of frequency spectrum obtained after substituting the masses as well as their positions of the middle column into the forward problem. The desired squared frequencies are in bold face in the right-hand column vectors to make it easier for the reader to compare them with the desired frequencies of the left-hand column.

From Table 7, some observations can be made. First, the desired value and order of the two input frequencies were preserved by the inverse problem. Specifically, in Table 7, the two input frequencies remain as the second and fourth system frequencies when the complete frequency spectrum was found for all three possible solutions returned by the inverse problem. Second, for each mass obtained, there must be two corresponding positions that are symmetrical with respect to the middle of the beam. This follows from the symmetry of the boundary conditions of the simply supported beam. Finally, a good equation solver is imperative for this method to work properly. For example, the use of an alternative equation solver (*DirectSearch*) yielded additional parameters c and p that were not returned by the built-in equation solver of Maple.

In order to verify that the method works when the chosen natural frequencies are not close together, a second example with the simply supported beam is presented. In this second example, the chosen frequencies are the second and eighth natural frequencies. As shown in Table 8, the method worked well in this case. Interestingly, a larger number of possible solutions were obtained in this case, in comparison with the cases shown in Tables 6 and 7. It is observed that the larger the separation in chosen order between λ_i and λ_j, the greater the number of possible solutions to the inverse problem.

Table 8. Inverse problem for second and eighth frequencies for a simply supported beam.

Input given to the inverse problem	Solution via inverse determinant method	Full span of frequency squared spectrum obtained via the solution of forward problem
$\lambda_2 = 1411.900275$, $\lambda_8 = 3.773881670 \times 10^5$ (Obtained with $m = 0.2\rho L$ $l = 0.4L$ in the forward code)	$m = 0.2\rho L$ $l = 0.6L$	71.3527 **1411.9003** 7164.65111 20420.52363 60880.6819 1.05946×10^5 2.21752×10^5 **3.77388×10^5** 5.73630×10^5 9.74091×10^5
	$m = 0.0807\rho L$ $l = 0.6615L$	86.6441 **1411.9003** 7887.5523 22430.0485 56495.1732 1.26101×10^5 2.13015×10^5 **3.77388×10^5** 6.37810×10^5 9.05601×10^5
	$m = 0.0752\rho L$ $l = 0.8447L$	94.2653 **1411.9003** 6953.8085 22707.0792 58428.8453 1.25669×10^5 2.31928×10^5 **3.77388×10^5** 5.86872×10^5 9.10470×10^5

Input given to the inverse problem	Solution via inverse determinant method	Full span of frequency squared spectrum obtained via the solution of forward problem
	$m = 0.0585\rho L$	90.83562
	$l = 0.7133L$	**1411.9003**
		7750.6323
		24432.2348
		55584.5685
		1.20289×10^5
		2.33870×10^5
		$\mathbf{3.77388\times10^5}$
		5.98869×10^5
		9.64184×10^5
	$m = 0.0573\rho L$	93.4974
	$l = 0.7944L$	**1411.9003**
		7262.2914
		24320.8984
		60836.8705
		1.20776×10^5
		2.15464×10^5
		$\mathbf{3.77388\times10^5}$
		6.30690×10^5
		9.72121×10^5

8. AT A GLANCE

The determinant method was used to demonstrate how to impose two natural frequencies on a dynamical system consisting of a beam to which a single lumped mass is attached. In this method, the known (design) variables are the two natural frequencies and the unknown variables are the magnitude of the attached mass as well as its position along the beam. The proposed method is easy to code and can accommodate any kind of eigenfunction. It was shown that the expected values of the added mass and its position are recovered from the inverse problem, in addition to the unexpected possible solutions. Even for this simple problem, a unique solution does not exist. The non-uniqueness can be considered as a benefit from the point of view of design possibilities. An investigation of the inverse problem also reveals that the order of the frequencies in the hierarchy of the whole frequency spectrum is conserved. Ideally, the degree of discretization should be the same for both forward and inverse problem. Although not discussed here, a promising observation made by Mir Hosseini and Baddour [21] through numerical simulations was that using a lower degree of discretization for the inverse problem could still give acceptable results from the point of view of engineering design, especially when lower orders of frequencies are involved.

9. CONCISE SUMMARY

In this chapter, two direct methods of solving inverse eigenvalue problems were presented and examples of the methods given. Both methods are not restricted in the type of matrices with which they can be employed. In both cases, nth order polynomial equations are obtained when working with nth order matrices, which must be solved for the inverse problem. Hence, good polynomial solvers, capable of producing the multiple solutions expected of both methods, are required. In using these inverse eigenvalue methods, careful care must be taken in setting up the forward modelling problem because the choice of modelling approach and chosen functions have a large effect on the order and solvability of the ensuing inverse problem.

REFERENCES

1. Chu MT., "Inverse eigenvalue problems," SIAM Rev., 40, 1–39, 1998.
2. Chu MT. and Golub GH., Inverse Eigenvalue Problems: Theory, Algorithms, and Applications. USA: Oxford University Press, Oxford, 2005.
3. Gladwell GML., Inverse Problems in Vibration. Kluwer Academic Publishers, Dordrecht, 2004.
4. Hochstadt H., "On some inverse problems in matrix theory," Arch. Math., 18, 2, 201–207, 1967.
5. Hald OH., "Inverse eigenvalue problems for Jacobi matrices," Linear Algebra Appl., 14, 1, 63–85, 1976.
6. Boley D. and Golub G., "Inverse eigenvalue problems for band matrices," in WatsonG. Ed., Numerical Analysis, vol. 630. Berlin/Heidelberg: Springer, 1978, 23–31.
7. de Boor C. and Saff EB., "Finite sequences of orthogonal polynomials connected by a Jacobi matrix," Linear Algebra Appl., 75, 43–55, 1986.
8. Gladwell GML., "The Inverse Problem for the Vibrating Beam," Proc. R. Soc. Lond. Math. Phys. Sci., 393, 1805, 277–295, 1984.
9. Gladwell GML. and Willms NB., "A discrete Gel'fand-Levitan method for band-matrix inverse eigenvalue problems," Inverse Probl., 5, 2, 165–179, 1989.

10. Ram YM. and Elhay S., "An inverse eigenvalue problem for the symmetric tridiagonal quadratic pencil with application to damped oscillatory systems," SIAM J. Appl. Math., 56, 1, 232–244, 1996.

11. de Oliveira GN., "Matrices with prescribed entries and eigenvalues. I," Proc Am Math Soc, 37, 380–386, 1973.

12. Dias da Silva JA., "Matrices with prescribed entries and characteristic polynomial," Proc. Am. Math. Soc., 45, 1, 31–37, 1974.

13. Knapp AW., "Basic Algebra," in Basic Algebra, Boston: Birkhäuser, 2006, p. 219.

14. Chang FR. and Chen HC., "The generalized Cayley–Hamilton theorem for standard pencils," Syst. Control Lett., 18, 179–182, 1992.

15. Artin M., "Algebra," in Algebra, 2nd ed., Boston: Pearson Prentice Hall, 2011, p. 140.

16. Coolidge JL., Treatise on Algebraic Plane Curves. Mineola, New York: Dover Publications, 1959.

17. Moiseev S., "Universal derivative-free optimization method with quadratic convergence," arXiv e-print 1102.1347, 2011.

18. Meirovitch L., "Principles and Techniques of Vibrations," Upper Saddle River, NJ: Prentice Hall, 1996, 542–543.

19. Dumond P. and Baddour N., "Can a brace be used to control the frequencies of a plate?" SpringerPlus, 2, 1, 558, 2013.

20. Dumond P. and Baddour N., "Effects of a scalloped and rectangular brace on the modeshapes of a brace-plate system," Int. J. Mech. Eng. Mechatron., vol. 1, no. 1, pp. 1–8, 2012.

21. Mir Hosseini F. and Baddour N., "A structured approach to solve the inverse eigenvalue problem for a beam with added mass," Math. Probl. Eng., 2014, 2014.

22. Meirovitch L., Principles and techniques of vibrations. Upper Saddle River, NJ: Prentice Hall, 1997.

23. Cha PD., "A general approach to formulating the frequency equation for a beam carrying miscellaneous attachments," J. Sound Vib., 286, 4–5, 921–939.

03 EIGENVALUE PROBLEMS

1. INTRODUCTION: AN OVERVIEW

Vibration is an inherent feature seen in all mechanical systems. Oscillation circuits in electrical systems display analog phenomena. Neglecting the consequences of vibration can lead to resonance, which can have both catastrophic consequences, such as the failure of a bridge, as well as advantageous uses. The vibratory phenomenon can be precisely characterized by the use of a differential equation or a collection of differential equations. Subsequently, this problem is transformed into the eigenvalue problem. The incentive for studying eigenvalue problems is the reason. The organization of this chapter is as follows: The subjects of interest encompass the Linear eigenvalue issue and the quadratic eigenvalue problem.

In natural sciences and engineering, are often used differential equations and systems of differential equations. Their solution leads to the problem of eigenvalues. Because of that, problem of eigenvalues occupies an important place in linear algebra. In this caption we will consider the problem of eigenvalues, and to linear and quadratic problems of eigenvalues. During the studying of linear problem of eigenvalues, we put emphasis on QR algorithm for unsymmetrical case and on minmax characterization of symmetric case. During the studying of quadratic problems of eingenvalue, we consider the linearization and variational characterization. We illustrate all with practical examples.

2. THE LINEAR EIGENVALUE PROBLEM

This section considers the linear eigenvalue problem of finding parameter λ such that the linear system

$$Ax=\lambda x \qquad (1)$$

has nontrivial solution **x**, where $\mathbf{A} \in \mathbf{C}^{(n,n)}$. The scalar λ is called **an eigenvalue** of **A,** and **x** is an **eigenvector** of **A corresponding** to λ. The set of all eigenvalues of matrix **A** is called the spectrum of **A**, and is denoted as $\sigma(\mathbf{A})$

The literature discusses the right and left eigenvectors. In our deliberations, we have been limited to the right eigenvectors, which are earlier defined.

This section is organized as follows:

- Basis properties (characteristic polynomial, bases for eigenspaces, eigenvalues and invertibility, diagonalization)
- QR Algorithm (The QR algorithm is used for determining all the eigenvalues of a matrix. Today, it is the best method for solving the unsymmetrical eigenvalue problems.)
- Mathematical background for Hermitian (symmetric) case (Rayleigh quotient, min max principle of Poincare, minmax principle of Courant-Fischer)
- Physical Background
- General linear eigenvalue problem

$$Ax=\lambda Bx \qquad (2)$$

2.1. Basic properties

In this section we outline the basic concepts and theorems, which will allow us to understand further elaboration. The eigenvalue problem is related to the homogeneous system of linear equations, as we will see in the following discussion.

To find the eigenvalues of $n \times n$ matrix **A** we rewrite (1) as

$$Ax=\lambda Ix \qquad (3)$$

or by inserting an identity matrix I equivalently

$$(A-\lambda I)x=0 \qquad (4)$$

Such system is homogeneous system and the system (3) has a nontrivial solution if and only if

$$\det(A-\lambda I)=0 \qquad (58)$$

This is called the **characteristic equation** of **A;** the scalars satisfying this equation are the eigenvalues of **A**. When is expanded, the determinant $\det(\mathbf{A} - \lambda\mathbf{I})$ is polynomial p in λ, and it is called the **characteristic polynomial** of **A**.

The following theorem gives the link between the characteristic polynomial of the matrix **A** and its eigenvalues.

Theorem 2.1.1. Equivalent Statements: If **A** is an $n \times n$ matrix and λ is a complex number, then the following are equivalent

- λ is an eigenvalue of **A**.
- The system of equations $(\mathbf{A} - \lambda\mathbf{I})\mathbf{x} = \mathbf{0}$ has nontrivial solutions.
- There is a nonzero vector $\mathbf{x}$ in $\mathbb{C}^n$ such that $\mathbf{Ax} = \lambda\mathbf{x}$.
- λ is a solution of the characteristic equation $det(\mathbf{A} - \lambda\mathbf{I})$.

Some coefficients of the characteristic polynomial of **A** have a specific shape. The following theorem gives the information about it.

Theorem 2.1.2.: If **A** is an $n \times n$ matrix, then the characteristic polynomial $p(\lambda)$ of **A** has degree n, the coefficient of λ^n is $(-1)^n$, the coefficient of λ^{n-1} is $(-1)^{n-1}$ $trace(\mathbf{A})$ and the constant term is $det(\mathbf{A})$, where $trace(\mathbf{A}) := a_{11} + a_{22} + \cdots + a_{nn}$.

In some structured matrices, eigenvalues can be read as shown in Theorem 2.1.3.

Theorem 2.1.3.: If **A** is an $n \times n$ triangular matrix (upper triangular, lower triangular, or diagonal), then the eigenvalues of **A** are entries of the main diagonal of **A**.

Cayley-Hamilton's theorem is one of the most important statements in linear algebra. The theorem states:

Theorem 2.1.4.: Substituting the matrix **A** for λ in characteristic polynomial of **A**, we get the result of zero matrix i.e., $p(\mathbf{A}) = 0$.

There are a number of methods for determining eigenvalue. Some methods allow finding all the eigenvalues and the other just a few of the eigenvalues. Methods based on first determining the coefficients of the characteristic polynomial, and later to determining the eigenvalue solving algebraic equations are rarely implemented, because they are numerically unstable. In fact, for the coefficients of the characteristic polynomial burdened with rounding errors, and due to numerical instability cause large errors in the eigenvalue. Because of that, the characteristic polynomial has mainly theoretical significance. The methods, which are based on the direct application characteristic polynomial, are applied in practice only when the characteristic polynomial is well conditioned. Also for some structured matrices, we can apply the method for the characteristic polynomial, but we don't calculate directly characteristic polynomial coefficients. The following example describes a class of such matrices.

Example 2.1.1 The Example of structured matrix which achieve the characteristic polynomial for determining the eigenvalue are Toeplitz matrix. Toeplitz matrix marked as $\mathbf{T_n}$, are matrices with constant diagonals. If the Toeplitz matrix is symmetric and positive definite, recursive relation is $p_n(\lambda) = p_{n-1}(\lambda)\beta_{n-1}(\lambda)$, where are p_n i p_{n-1} characteristic polynomial matrix $\mathbf{T_n}$ i $\mathbf{T_{n-1}}$ respectively a β_{n-1} Schur-Szegö parameter for Yule-Walker system. The above recursive relation enables work with characteristic polynomial without individual accounts of his odds. More information can be found at [1]

The following definitions are introducing two important terms: the geometric multiplicity of λ_0 and the algebraic multiplicity of **A**.

The eigenvectors corresponding to λ are the nonzero vectors in the solutions space of $(\mathbf{A} - \lambda\mathbf{I})\mathbf{x} = \mathbf{0}$. We call this solution space the **eigenspace** of **A** corresponding to λ.

Definition 2.1.1.: If λ_0 is an eigenvalue of an $n \times n$ matrix **A**, then the dimension of the eigenspace corresponding to λ_0 is called the **geometric multiplicity** of λ_0, and the number of times that $\lambda - \lambda_0$ appears as a factor in the characteristic polynomial of **A** is called the **algebraic multiplicity** of **A**.

Eigenvalue and eigenvector have some specific features, which are easy to prove.

The following are some of the obvious features of eigenvalues of matrix A and corresponding eigenvector:

Theorem 2.1.5

- If $\mu \neq 0$ complex number, λ is an eigenvalue of matrix **A**, and $\mathbf{x} \neq \mathbf{0}$ corresponding eigenvector, then $\mu\mathbf{x}$ is a corresponding eigenvector.
- If k is a positive integer, λ is an eigenvalue of matrix **A**, and $\mathbf{x} \neq \mathbf{0}$ corresponding eigenvector, then λ^k is an eigenvalue of $\mathbf{A}^k$ and **x** is a corresponding eigenvector.
- Matrix **A** and $\mathbf{A}^T$ have the same eigenvalues.

In linear algebra invertible matrix are important. From the problem of eigenvalues we can easily conclude If the matrix **A** is invertible or not. What more can be, the eigenvalues of the matrix **A** invertible can be immediately read from the eigenvalues of the matrix A^{-1}. Because of that, in the following theorems we summarize some properties of invertible matrix.

Theorem 2.1.6.: If **A** is an $n \times n$ matrix, then the following are equivalent.

- **A** is invertible.
- $\lambda = 0$ is not an eigenvalue of **A**
- If λ is an eigenvalue of matrix invertible **A**, and $\mathbf{x} \neq \mathbf{0}$ corresponding eigenvector, then $\frac{1}{\lambda}$ is an eigenvalue of $\mathbf{A}^{-1}$ and **x** is a corresponding eigenvector.
- $det(\mathbf{A}) \neq 0$.
- **A** has rank n.
- $\mathbf{Ax} = \mathbf{0}$ has only the trivial solution.
- $\mathbf{Ax} = \mathbf{b}$ has exactly one solution for every $n \times 1$ matrix B

- The column vectors of **A** form a basis for $\mathbb{R}^n$.
- The row vectors of **A** form a basis for $\mathbb{R}^n$.
- $\mathbf{A}^T\mathbf{A}$ is invertible.

The problem of finding a base $\mathbb{R}^n$ consisting of eigenvectors is very important in linear algebra. Because of that, in this section we will consider the following two equivalent problems:

The Eigenvector Problem. Given an $n \times n$ matrix **A**, does there exist a basis for $\mathbb{R}^n$ consisting of eigenvectors of **A**?

The Diagonalization Problem (Matrix Form). Given an $n \times n$ matrix **A**, does there exist an invertible matrix **P** such that $\mathbf{P}^{-1}\mathbf{AP}$ is a diagonal matrix?

The latter problem suggests the following terminology.

Definition 2.1.3: Two square matrices **A** and **B** are similarly called, if there is invertible matrix **P**, so that $\mathbf{B} = \mathbf{P}^{-1}\mathbf{AP}$. The transition from **A** to the matrix of the matrix **B** is called the similarity transformations .

The importance of similar matrices can be seen in the following theorem:

Theorem 2.1.7: Similar matrices **A** and **B** have the same characteristic polynomial. They have the same eigenvalues including their geometric multiplicity of the geometric multiplicity of λ_0 and the algebraic multiplicity of λ_0.
Based on previous definitions, we can define the term diagonalizable as follows:

Definition 2.1.4: A square matrix **A** is called diagonalizable if the transformation of similarity may be translated into a diagonal form. To the above two problems are obviously equivalent to the following theorem.

Theorem 2.1.8: If **A** is an $n \times n$ matrix, then the following are equivalent.

- **A** is diagonalizable.
- **A** has n linearly independent eigenvectors.

The following algorithm is for diagonalizing a matrix

Algorithm for Diagonalizing a Matrix

Find n linearly independent eigenvectors of **A**, marked as $\mathbf{x}^1, \mathbf{x}^2, \cdots, \mathbf{x}^n$. Form the matrix **P** having $\mathbf{x}^1, \mathbf{x}^2, \cdots, \mathbf{x}^n$., at its column vectors. The matrix $\mathbf{P}^{-1}\mathbf{AP}$ then will be diagonal with $\lambda_1, \lambda_2, \cdots, \lambda_n$ successive diagonal entries, where λ_i is eigenvalue corresponding to $\mathbf{x}^i$, for $i = 1, 2, \cdots, n$.

Theorem 2.1.9: If $\mathbf{x}^1, \mathbf{x}^2, \cdots, \mathbf{x}^k$ are eigenvectors of **A** corresponding to distinct eigenvalues $\lambda_1, \lambda_2, \cdots, \lambda_k$, then $\{\mathbf{x}^1, \mathbf{x}^2, \cdots, \mathbf{x}^k\}$ is a linearly independent set. As consequence of Theorem 2.1.8. and Theorem 2.1.9 ., we obtain the following important result

Theorem 2.1.10: If an $n \times n$ matrix **A** has n distinct eigenvalues, then **A** is diagonalizable. There are matrices that can have the same eigenvalues and yet can be diagonalizable. Broadest such class of such matrices are normal matrix, which we will introduce the following definition.

Definition 2.1.5 . Matrix $\mathbf{A} \in \mathbb{C}^{(n,n)}$ is called normal, if holds $\mathbf{A}^H\mathbf{A} = \mathbf{A}\mathbf{A}^H$.

More general characterization of diagonalizable matrix A is given in the following theorem .

Theorem 2.1.11: If **A** is a square matrix, then:

- For every eigenvalue of **A** the geometric multiplicity is less than or equal to the algebraic multiplicity.
- **A** is diagonalizable if and if the geometric multiplicity is equal to the algebraic multiplicity for every eigenvalue.

2.2. QR algorithm

In this section we will present the QR algorithm. In numerical linear algebra, the **QR algorithm** is an eigenvalue algorithm : that is, a procedure to calculate the eigenvalues and eigenvectors of a matrix. QR algorithm represent factorization method, because it is based on the matrix decomposition. First factorization method labeled as LR algorithm is developed by H. Rutishauser in 1958. LR algorithm is due to the many shortcomings, today is rarely used, (Wilkinson monograph). Better factorization method, is designated as the QR algorithm. The basic form have developed independently from each other, in 1962 G. F. Francis (England) and by Vera N. Kublanskoovskaya (USSR). Today, it is the best method for solving the unsymmetrical eigenvalue problems, when you want to determine all the eigenvalues of the matrix. It is particularly effective when it is brought into the so-called matrix "Condensed form". Condensed form is for the unsymmetrical problem Hessenberg form. About Hessenberg form will be discussed more later. The basic idea is to perform a QR decomposition, writing the matrix as a product of an orthogonal matrix and an upper triangular matrix, multiply the factors in the reverse order, and iterate.

For understanding of the QR algorithm we will need the following terms:

Definition 2.2.1. Matrix $\mathbf{Q} \in P^{(n,n)}$ called orthogonal if worth $\mathbf{Q}^T\mathbf{Q} = \mathbf{I}$.

Remark 2.2.1. Orthogonal matrix represent special case of unitary matrices. The Matrix $\mathbf{U} \in \mathbb{C}^{(n,n)}$ called unitary if applies $\mathbf{U}^H\mathbf{U} = \mathbf{I}$. It is now clear that the orthogonal matrix of a unitary matrix in which all elements of real numbers. In practice, of course, is easier to work with the orthogonal than unitary matrices.

Remark 2.2.2 Orthogonal and unitary matrices are normal matrices.

Let **A** and **B** are similar matrices. If the similarity transformations performed by the orthogonal or unitary matrix **Q** i.e. if applies $\mathbf{B} = \mathbf{Q}^T\mathbf{AQ}$ or $\mathbf{B} = \mathbf{U}^H\mathbf{AU}$ we will say that the matrices **A** and **B** are unitary similar. Since the unitary similar matrices are special special case of similar matrix, the eigenvalues of unitary similar matrices are the same.

In addition to the unitary similar matrices and their properties for the introduction of the QR algorithm, we will need the following theorem.

Theorem 2.2.1. Let's $\mathbf{A} \in \mathbb{R}^{(n,n)}$ regular matrix, then there is a decomposition $\mathbf{A} = \mathbf{QR}$, where is **Q** orthogonal matrix and **R** upper triangular matrix. If the diagonal elements of the matrix **R** are positive, decomposition is unique.

The decomposition of the matrix **A** from Theorem 2.2.1 is called the QR decomposition of the matrix **A**.

The following is a basic form of the QR algorithm.

Let $\mathbf{A}_0 := \mathbf{A}$. The basic forms of the QR algorithm is given by

Algorithm 2.2.1 . (QR algorithm-basic forms)

For $i = 0, 1, \cdots$ until convergence

Decompose $\mathbf{A}_i = \mathbf{Q}_i\mathbf{R}_i$ (QR decomposition)

$$A_{i+1} = R_i Q_i 1$$

End

Theorem 2.2.1 All matrices $\mathbf{A}_i$ resulting in algorithms 2.2.1 are unitary similar.

Proof

Since applies $Q_i^T A_i = R_i$ we have $A_{i+1} = R_i Q_i = Q_i^T A_i Q_i$. Based on past relationships, it is clear that it is true $A_{i+1} = Q_i^T Q_{i-1}^T \cdot \cdots \cdot Q_0^T A_0 Q_0 \cdot \cdots \cdot Q_{i-1} Q_i$. It is obvious that the matrix $Q := Q_0 \cdot \cdots \cdot Q_i$ orthogonal matrix and the theorem is proved.

Let's look briefly at some properties QR algorithm:

- For QR factorization full matrix it is necessary $O(n^3)$ flops by factorization. Total QR iteration needs $O(n^4)$ flops, which means that it is objectively very slow. But the problem of the slowness of the basic forms of the algorithm can be overcome with the two strategies, namely: translation matrix in Hessenberg form and the introduction of shift.

Let $\mathbf{A} \in \mathbb{R}^{(n,n)}$ then for each and $\mathbf{Q}_i, \mathbf{R}_i \in \mathbb{R}^{(n,n)}$ and the whole algorithm is performed in real area. If all eigenvalues of matrix A modulo different, then they are all real and the algorithm converges. However, if there is at least one, and so it is worth $\left|\frac{\lambda_{i+1}}{\lambda_i}\right| \approx 1$ QR algorithm is very slow. If the matrix **A** has complex eigenvalues then a basic form of the QR algorithm does not converge. We have already seen the need for the introduction of Hessenberg form of a matrix and it will introduce the following definition.

Definition 2.2.2. For the matrix **A** will say that in the upper Hessenberg form if holds

$$a_{ij}=0 \text{ for } i\text{-}j>2$$

i.e. the upper triangular matrix has a side below the main diagonal.

The process of reducing the Hessenberg form can be performed using Househelder reflectors and Givens rotation. Let's look at bringing in Hessenberg form using Househelder spotlight.

$$\text{Let } A=\begin{pmatrix} a_{11} & \mathbf{c}^T \\ \mathbf{b} & B \end{pmatrix} \text{ with } \mathbf{b}\neq 0$$

Our goal is to determine $\omega \in \mathbb{C}^{n-1}$ with features a$\|\omega\|_2 = 1$ and

$$Q_1\mathbf{b} := (I_{n-1}-2\omega\omega^H)\mathbf{b}=k\mathbf{e}^1$$

where $\mathbf{I}_{n-1}$ is identity matrix row *n-1* a $\mathbf{e}^1$ the first column of the matrix $\mathbf{I}_{n-1}$.

We define Householder reflector as follows

$$P_1 := \begin{pmatrix} 1 & 0^T \\ 0 & Q_1 \end{pmatrix}$$

Now applies

$$A_1 := P_1AP_1 = \begin{pmatrix} a_{11} & \mathbf{c}^T Q_1 \\ k & \\ 0 & \\ \vdots & Q_1BQ_1 \\ 0 & \end{pmatrix}$$

Obviously, the first column of the matrix $\mathbf{A}_1$ have a look what is required in the upper Hessenberg form. In this way, we showed that the first column is converted into a suitable form. An analogous procedure can be performed on 2,3,...,*n-1* column.

For the implementation of QR algorithm is important that in each iteration in implementing QR algorithm preserves the structure of the matrix. For a matrix form in the upper Hessenberg applies following theorem.

Theorem 2.2.2 If *A* is upper Hessenberg matrix, then the **Q** in its QR factorization **A=QR** also upper Hessenberg matrix.

The above theorem that QR and also upper Hessenberg matrix is the product of the upper triangular matrix and upper Hessenberg matrix. Preserving the structure of the matrix is very important to stop the efficiency of the algorithm. Namely, if the upper Hessenberg matrix **A** matrix of its QR factorization is only $O(n^2)$ instead of $O(n^3)$ as is necessary for QR factorization (decomposition) of full matrix.

Let's look at one more advantage of the transformation matrix in Hessenberg form. Namely, $\mathbf{A}_i \rightarrow \mathbf{R}$ $(i \rightarrow \infty)$, where **R** is an upper triangular matrix. Because the matrix $\mathbf{A}_i$ is always upper Hessenberg matrix proves that elements of the second diagonal tend to zero and it's worth $a^{(i)}_{j+1,j} \rightarrow 0\ (i \rightarrow \infty),\ j = 1, 2, \cdots, n-1$ wherein $a^{(i)}_{j+1,j}$ elements of the matrix $\mathbf{A}_i$. It is now clear that for sufficiently large and can be read eigenvalues initial value of A on the basis of theorem 2.1.3. as the diagonal elements of the matrix $\mathbf{A}_i$

For a further improve of the algorithm, shift is used . The idea of improving the QR algorithm is based on the simple fact that if the eigenvalues of **A** are equal λ_i that the eigenvalues of the matrix $\mathbf{A} - \sigma\mathbf{I}$ equal $\lambda_i - \sigma$. If we shift σ chosen close eigenvalues, there is a strong acceleration of the algorithm.

Let $\mathbf{A}_0 := \mathbf{A}$.

Algorithm 2.2.2. . (QR algorithm with shift)

For $i = 0, 1, \cdots$ until convergence

Choose shift σ_i near the eigenvalues value

Decompose $\mathbf{A}_i - \sigma_i\mathbf{I} = \mathbf{Q}_i\mathbf{R}_i$ (QR decomposition)

$$A_{i+1} = R_i Q_i + \sigma_i I$$

End

It is easy to prove that all matrices in the algorithm 2.2.1. are unitary similar. From the above it is clear that in the case of real matrices with real eigenvalues is best to shift the parameters taken $\sigma_i = a^{(i)}_{n.n}$..

For a further analysis of the parameter shifts we will need the concept of reduced upper Hessenberg matrix as well as the implicit Q theorem.

Definition 2.2.3 The upper Hessenberg matrix H is **unreduced upper Hessenberg matrix** if the first second diagonal not a single zero.

Theorem 2.2.3. Let $\mathbf{Q}^T\mathbf{A}\mathbf{Q} = \mathbf{H}$ unreduced upper Hessenberg matrix with positive subdiagonal elements $h_{k+1,k}$ a **Q** unitary matrix. The columns of the matrix **Q** and matrix **H** starting from the second to n-th, are uniquely determined first column of the matrix **Q**.

Proof

Let $\mathbf{Q} = (\mathbf{q}^1, \mathbf{q}^2, \cdots, \mathbf{q}^n)$ some prior $\mathbf{q}^1, \mathbf{q}^2, \cdots, \mathbf{q}^k$ and the first $k - 1$ column of the matrix H is determined. The proof will be carried out by mathematical induction by k. For

$k = 1$ is determined $\mathbf{q}^1$ and the process can start. Because $\mathbf{QH} = \mathbf{AQ}$ and $\mathbf{H} = (h_{ij})$ the upper Hessenberg matrix applies

$$h_{k+1,k}\boldsymbol{q}^{k+1} + h_{kk}\boldsymbol{q}^{k} + \cdots + h_{1k}\boldsymbol{q}^{1} = A\boldsymbol{q}^{k}$$

If you multiply the last equality with $(\mathbf{q}^i)^H$ we get

$$h_{ik} = \left(q^i\right)^H Aq^k \left(i = 1, 2, \cdots, k\right)$$

From here it's k-th column of H except element $h_{k+1,k}$ specified.

Because $h_{k+1,k} \neq 0$ we have

$$q^{k+1} = \frac{1}{h_{k+1,k}}\left(Aq^k - \sum_{i=1}^{k} h_{ik}\, q^i\right)$$

From $(\mathbf{q}^{k+1})^H\mathbf{q}^{k+1} = 1$ and positivity $h_{k+1,k}$ we get $h_{k+1,k}$ in a unitary way.

Theorem is proved.

Remark 2.2.3 Condition $h_{k+1,k} > 0$ in the previous theorem was we only need to ensure uniformity of matrices **Q** i **H.**

With the help of implicit Q theorem we discuss the selection of shift $\mathbf{A} = \mathbf{A}_0$ real matrix has complex eigenvalues. Then you have to make a double shift for σ i $\bar{\sigma}$. Namely,

$$A_0 - \sigma \boldsymbol{I} = \boldsymbol{Q}_1\boldsymbol{R}_1, \quad A_1 = \boldsymbol{R}_1\boldsymbol{Q}_1 + \sigma \boldsymbol{I}$$
$$A_0 - \bar{\sigma} \boldsymbol{I} = \boldsymbol{Q}_1\boldsymbol{R}_1, \quad A_1 = \boldsymbol{R}_1\boldsymbol{Q}_1 + \bar{\sigma} \boldsymbol{I}.$$

From there, easy to get $A_2 = Q_2^T Q_1^T A_0 Q_1 Q_2$ Matrices $\mathbf{Q}_1$ i $\mathbf{Q}_2$ can be chosen so that $\mathbf{Q}_1\mathbf{Q}_2$ real matrices and therefore the matrix $\mathbf{A}_2$ is real matrix. Applying

$$Q_1Q_2R_2R_1 = Q_1\left(A_1 - \bar{\sigma}I\right)R_1 = Q_1\left(R_1Q_1 + (\sigma - \bar{\sigma})I\right)R_1 = Q_1R_1Q_1R_1 + (\sigma - \bar{\sigma})Q_1R_1$$
$$= \left(A_0 - \sigma I\right)^2 + (\sigma - \bar{\sigma})\left(A_0 - \sigma I\right) = A_0^2 + \left(\sigma + \bar{\sigma}\right)A_0 + \left|\sigma\right|^2 I =: M.$$

Because of $\sigma + \bar{\sigma} \in P$, matrix M is real. Then $\mathbf{Q}_1\mathbf{Q}_2\mathbf{R}_2\mathbf{R}_1$ QR factorization real matrix and that means $\mathbf{Q}_1\mathbf{Q}_2$ i $\mathbf{R}_2\mathbf{R}_1$ we can choose a real matrix. The first column of the matrix $\mathbf{Q}_1\mathbf{Q}_2$ is proportional to the first column of the matrix M, and the other columns are calculated by applying implicit Q theorem.

2.3. Mathematical background for Hermitian (symmetric) case

In this section we look at the problem of eigenvalues in the case of a symmetric or Hermitian matrix.

Definition 2.3.1. Matrix $\mathbf{A} \in \mathbb{R}^{(n,n)}$ is called **symmetric** if applies $\mathbf{A} = \mathbf{A}^T$.

Definition 2.3.2 Matrix $\mathbf{A} \in \mathbb{C}^{(n,n)}$ is called **Hermitian** if applies $\mathbf{A} = \mathbf{A}^H$.

Remark 2.3.1 Symmetric Matrices are only a special case of Hermitian matrices in the case that the elements matrices are real numbers. Therefore, we will formulate a theorem for Hermitian matrices.

Remark 2.3.2 Hermitian and symmetric matrices are normal matrices, which means that they can diagonalize.

The following theorem gives important information on the reality of eigenvalues Hermitian (symmetric) matrices. This feature greatly facilitates consideration of the problem of eigenvalues for this class of matrices, which makes this class of matrices applicable in practice.

Theorem 2.3.1 . If **A** is Hermitian (symmetric) matrix, then:

- The eigenvalues of **A** are all real numbers.
- Eigenvectors from different eigenspace are orthogonal.

Since all the eigenvalues of real ones can be compared. Therefore, we assume that the eigenvalues in order of size, i.e. It applies $\lambda_1 \leq \lambda_2 \leq \cdots \leq \lambda_n$ and that the corresponding orthonormal eigenvectors.

If the matrix **A** is symmetrical due to symmetry feature, it comes to the significant acceleration algorithms for the unsymmetrical case. We will demonstrate the QR algorithm which is presented in section 2.2. for the unsymmetrical case. In symmetric case is important to note that the upper Hessenberg form of symmetric matrix tridiagonal matrix whose QR decomposition is only necessary $O(n)$ operation. It is also important that during the QR algorithm preserves the structure of the matrix or all tridiagonal matrices $\mathbf{A}_i$. For a shift in this case is usually taken Wilkins shift which is defined as an eigenvalue of matrix

$$\begin{pmatrix} a_{n-1,n-1} & a_{n-1,n} \\ a_{n-1,n} & a_{n,n} \end{pmatrix}, \text{ that is closest } a_{n,n}$$

For a QR algorithm Wilkinson shifts apply following theorem, whose proof is given in [2]

Theorem 2.3.2 (Wilkinson) QR algorithm with Wilkinson shifts for symmetric tridiagonal matrix converges globally and at least linearly. For a almost all of the matrices are on asymptotically cubic converging.

Now we introduce the very important concept of the Rayleigh quotient, because it gives the best estimate of the eigenvalues for a given vector $\mathbf{x} \in \mathbb{C}^n$, $\mathbf{x} \neq 0$

Definition 2.3.3 Let A be a Hermitian (symmetric) matrix. For a given vector $\mathbf{x} \in \mathbb{C}^n$, $\mathbf{x} \neq 0$ **Rayleigh quotient** is defined $R(x) := \frac{\mathrm{x}^{\mathrm{H}} A \mathrm{x}}{\mathrm{x}^{\mathrm{H}} \mathrm{x}}$.

The importance of the Rayleigh quotient is seen in the following theorems

Theorem 2.3.3. (Features Rayleigh quotient)

- For all $\mathbf{x} \in \mathbb{C}^n$, $\mathbf{x} \neq \mathbf{0}$ worth $\lambda_1 \leq R(\mathbf{x}) \leq \lambda_n$.
- $\lambda 1=\min_{x\neq 0}R(x),\lambda_n=\max_{x\neq 0}R(x)$
- If $\mathbf{x} \neq \mathbf{0}$ with $\lambda_1 = R(\mathbf{x})$ respectively $\lambda_n = R(\mathbf{x})$ $\mathbf{x}$ **is** then x is the eigenvector corresponding to λ_1 respectively λ_n.
- $\lambda i = \min\{R(x) : x^H x^j = 0, j = 1, \cdots, i-1, x \neq 0\}$ $=\max\{R(x):x^H x^j=0,j=i+1,\cdots n,x\neq 0\}$

Paragraph (d) in the previous theorem is known as Rayleigh principle. However it is numerically worthless, because for the determination example, we need eigenvector or $\mathbf{x}^1$ corresponding to the eigenvalue λ_1. In order to overcome this disadvantage is introduced min max principle of Poincaré that is listed in the following theorems.

Theorem 2.3.4 . (min max principle of Poincaré)

$$\lambda_i = \min_{dimV=i} \max_{\mathbf{x}\in V\setminus\{\mathbf{0}\}} R(\mathbf{x}) = \max_{dimV=n-i+1} \min_{\mathbf{x}\in V\setminus\{\mathbf{0}\}} R(\mathbf{x})$$

The following formulation is known as min max principle of Courant-Fischer and often favorable to use.

Theorem 2.3.5. (min max principle of Courant-Fischer)

$$\lambda_i = \min_{\{p^1,\cdots,p^{i-1}\}} \max\left\{R(\mathbf{x}):\mathbf{x}^{\mathrm{H}}\mathbf{p}^j = 0,\ j=i+1,\cdots n,\mathbf{x}\neq\mathbf{0}\right\}$$

$$\max_{\{p^1,\cdots,p^{n-i}\}} \min\left\{R(\mathbf{x}):\mathbf{x}^{\mathrm{H}}\mathbf{p}^j = 0,\ j=i+1,\cdots n,\mathbf{x}\neq\mathbf{0}\right\}.$$

From the above it is clear that these theorems are important for the localization of eigenvalues.

The following is an algorithm that is in linear algebra known as Rayleigh quotient iteration and it reads as follows

Let $\mathbf{A} \in \mathbb{R}^{(n,n)}$ symmetric matrix and $\mathbf{x}^0$ initial vector, which is the standardized. For which applies $||\mathbf{x}^0||_2 = 1$.

Algorithm 2.3.1 . (Rayleigh quotient iteration)

$$\sigma_0=(x^0)^T A x^0$$

For $i = 1, 2 \cdots$ until convergence

Solve $(A_i - \sigma_i \mathbf{I})\mathbf{y}^i = \mathbf{x}^{i-1}$

$$\mathbf{x}^i = \frac{\mathbf{y}^i}{\mathbf{y}^i_2}$$

$$\sigma_0 = \left(\mathbf{x}^i\right)^T \mathbf{A}\mathbf{x}^i.$$

End

Theorem 2.3.6. Rayleigh quotient iteration converges cubic.

Finally point out that the most effective for symmetric matrices Divide-and-Conquer method. This method was introduced by Cupen [3] and the first effective implementation is the work of Cu and Eisenstat [4]. About this method, more information can be found in [4].

2.4. Physical background

We have already mentioned that the problem of eigenvalues has numerous applications in engineering. Even more motivation to consider problems of eigenvalues, comes from their large application in technical disciplines. On a simple example of a mass-spring system illustrated with application of eigenvalues in engineering. We are assuming, that each spring has the same natural length l and the same spring constant k. Finally, we can assume that the displacement of each spring in measured to its own local coordinate system with an origin at the spring's equilibrium position.

Applying Newton's second law we get the following system

$$m_1 \frac{d^2 x_1}{dt^2} - k\left(-2x_1 + x_2\right) = 0$$

$$m_2 \frac{d^2 x_2}{dt^2} - k\left(x_1 - 2x_2\right) = 0.$$

We are aware of vibration theory that

$$x_i = a_i \sin \omega t,$$

where a_i is the amplitude of the vibration of mass i and ω is the frequency of the vibration.

If the last equation twice differentiate by t, we get

$$\frac{d^2 x_2}{dt^2} = -a_i \omega^2 \sin \omega t.$$

If the last two equations obtained expressions for x_i i $\frac{d^2x_i}{dt^2}$ $(i=1, 2)$ replace the initial system and write the system in matrix form, then we get

$$A\begin{pmatrix} a_1 \\ a_2 \end{pmatrix} = \omega^2\begin{pmatrix} a_1 \\ a_2 \end{pmatrix},$$

where

$$A = \begin{pmatrix} \frac{2k}{m_1} & -\frac{k}{m_1} \\ -\frac{k}{m_2} & \frac{2k}{m_2} \end{pmatrix}$$

Equation (6) represents unsymmetrical eigenvalue problem.

More information on this case can be found in [5]

2.5. General Linear Eigenvalue Problem

In this section we will deal with general linear eigenvalue problem or the problem

$$\mathbf{Ax} = \mathbf{Bx}, \quad \mathbf{x} \neq \mathbf{0},$$

where **A, B** $\in C^{(n,n)}$

The scalar λ is called an eigenvalue of the problem (7), and **x** said to be an eigenvector of (7) corresponding to λ.

A common acronym for general linear eigenvalue problem is GEP. Now eigenvalue problems previously discussed is called the standard eigenvalue problem and tagging with SEP. In practice, the more often we meet with GEP than SEP. Now let's consider some features of GEP and establish its relationship with SEP.

It is obvious that the eigenvalues of (7) zero of the characteristic polynomial, which is defined as $p_n(\lambda) := det(\mathbf{A} - \lambda\mathbf{B})$. In the case of GES degree polynomial p_n is less than or equal to n. The characteristic polynomial of degree n has *pn* if and only if *B* is regular matrix. In case *B=I* we get SEP and in this case SEP has n eigenvalues. GEP and can be less than n eigenvalues. It can also happen that at GEP is $p_n(\lambda)\equiv 0$ and it that case GEP has infinitely many eigenvalues.

The following two examples illustrate this situation with GEP

Example 4.5.1 Let in GEP $A=\begin{pmatrix} 0 & 0 \\ 0 & 0 \end{pmatrix}$, $B=\begin{pmatrix} 0 & 6 \\ 0 & 0 \end{pmatrix}$. Then worth $p_n(\lambda)\equiv 0$ and every $\lambda \in \mathbb{C}$ is eigenvalue i $\mathrm{x}=\begin{pmatrix} 1 \\ 0 \end{pmatrix}$ is the corresponding eigenvector.

Example 4.5.2 Let in GEP $A=\begin{pmatrix}1 & 2\\ 3 & 4\end{pmatrix}$, $B=\begin{pmatrix}1 & -1\\ -1 & 1\end{pmatrix}$. When we get to the characteristic polynomial $p_n(\lambda)=(1-\lambda)(4-\lambda)-(3+\lambda)(2+\lambda)=-2-10\lambda$ and the only eigenvalue value is $\lambda=-\frac{1}{5}$=-15.

Atypical nature of we met in the last two examples are the result of the fact that in their matrix **B** was not regular. Therefore, it is usual in the GEP taken that **A** and **B** are Hermitian matrix and **B** is positive definite, that is, all the eigenvalues of the matrix **B** are positive.

Our goal is to find a connection between this taken GEP and symmetric SEP. As we said **B** is positive definite, **B** has a Cholesky decomposition, i.e., $\mathbf{B} = \mathbf{CC}^H$. Then is

$$A\mathbf{x}=\lambda B\mathbf{x} \Leftrightarrow Fy := C^{-1}AC^{-H}\mathbf{y}, \quad \mathbf{y} := C^H\mathbf{x}$$

Since matrix F has n eigenvalues and it belongs to the GEP also has n real eigenvalues. Namely on the basis of (7) are the same eigenvalues.

Let $\mathbf{y}^i$ i $\mathbf{y}^j$ orthonormal eigenvectors of F then for $\mathbf{x}^i = \mathbf{C}^{-H}\mathbf{y}^i$ **i** $\mathbf{x}^j = \mathbf{C}^{-H}\mathbf{y}^j$ apply

$$\begin{aligned}\delta_{ij} &= \left(\mathbf{y}^i\right)^H \mathbf{y}^j = \left(C^H\mathbf{x}^i\right)^H C^H\mathbf{x}^j \\ &= \left(\mathbf{x}^i\right)^H CC^H\mathbf{x}^j = \left(\mathbf{x}^i\right)^H Bx^j\end{aligned}$$

From equation (8) it is clear that the eigenvectors GEP (6) are an orthonormal vectors in relation to the new inner product is defined as $\langle \mathbf{x}, \mathbf{y}\rangle_B := \mathbf{x}^H\mathbf{B}\mathbf{y}$. Now we're going to GEP (7) redefine Rayleigh quotient as follows $R_{(A,B)}(x) := \frac{x^H Ax}{x^H Bx}$.. With this predefined inneren product and Rayleigh quotient applies to all theorems of section 2.3 with appropriate modification of the definition changes required. This gives self-generation Hermitian (symmetrical) SEP.

3. THE QUADRATIC EIGENVALUE PROBLEMS

In practice, nonlinear eigenproblems commonly arise in dynamic/stability analysis of structures and in fluid mechanics, electronic behavior of semiconductor hetero-structures, vibration of fluid–solid structures, vibration of sandwich plates, accelerator design, vibro-acoustics of piezoelectric/poroelestic structures, nonlinear integrated optics, regularization on total least squares problems and stability of delay differential equations. In practice, the most important is the quadratic problem $\mathbf{Q}(\lambda)\mathbf{x} = \mathbf{0}$ where

$$\mathbf{Q}(\lambda) := \lambda^2 A + \lambda B + C, \quad A,\ B,\ C \in \mathbb{C}^{n\times n}, \quad A\neq 0,\ \mathbf{x}\neq 0$$

This section is organized as follows:

- Basic Properties

We consider Rayleigh functional and Minmax Characterization

- Linearization

A standard approach for investigating or numerically solving quadratic eigenvalue linearization problems, where the original problem is transformed into a generalized linear eigenvalues problem with the same spectrum.

- Physical Background

We study vibration analysis of structural systems

3.1. Some Basic Properties

Variational characterization is important for finding eigenvalues. In this section we give a brief review of variational characterization of nonlinear eigenvalue problems. Since the quadratic eigenproblems are a special case of nonlinear eigenvalue problems, results for nonlinear eigenvalue problems can be specially applied for the quadratic eigenvalue problems. Variational characterization is generalization of well known minmax characterization for the linear eigenvalue problems.

We consider nonlinear eigenvalue problems

$$\boldsymbol{T}(\lambda)x = \mathbf{0},$$

where $\mathbf{T}(\lambda) \in \mathbb{C}^{n \times n}$, $\lambda \in J$ is a family of the Hermitian matrices depending continuously on the parameter $\lambda \in J$ and J is a real open interval which may be unbounded.

Problems of this type arise in damped vibrations of structures, conservative gyroscopic systems, lateral buckling problems, problems with retarded arguments, fluid–solid vibrations, and quantum dot heterostructures.

To generalize the variational characterization of eigenvalues we need a generalization of the Rayleigh quotient. To this end we assume that

(A) for every fixed $\mathbf{x} \in \mathbb{C}^n$, $\mathbf{x} \neq \mathbf{0}$ the scalar real equation

$$f(\lambda;\mathbf{x}) := \mathbf{x}^{\mathbf{H}}\boldsymbol{T}(\lambda)\mathbf{x}$$

has at most one solution $p(x) \in J$. Then $f(\lambda; \mathbf{x}) = 0$ implicitly defines a functional p on some subset $D \subset \mathbb{C}^n$ which is called the Rayleigh functional of (11).

(B) for every $x \in D$ and every $\lambda \in J$ with $\lambda \neq p(x)$ it holds that $(\lambda - p(x))f(\lambda; x) > 0$.

If p is defined on $D = \mathbb{C}^n\backslash\{\mathbf{0}\}$ then the problem (1) is called **overdamped**, otherwise it is called **nonoverdamped.**

Generalizations of the minmax and the maxmin characterizations of the eigenvalues were proved by Duffin [6] for the quadratic case and by Rogers [7] for the general overdamped problems. For the nonoverdamped eigenproblems the natural ordering to call the smallest eigenvalue the first one, the second smallest the second one, etc., is not appropriate. The next theorem is proved in [8], which gives more information about the following minmax characterization for eigenvalues.

Theorem 3.1.1: Let J be an open interval in $\mathbb{R}$, and let $\lambda) \in \mathbb{C}^{nxn}$, $\lambda \in J$, be a family of Hermitian matrices depending continuously on the parameter $\lambda \in J$, such that the conditions (A) and (B) are satisfied. Then the following statements hold.

For every $l \in \mathbb{N}$ there is at most one lth eigenvalue of $\mathbf{T}(\cdot)$ which can be characterized by

$$\lambda_l = \min_{V \in H_l, V \cap D \neq \varnothing} \sup_{v \in V \cap D} p(v)$$

If

$$\lambda_l := \inf_{V \in H_l, V \cap D \neq \varnothing} \sup_{v \in V \cap D} p(v) \in J.$$

For some $l \in \mathbb{N}$ then λ_l is the lth eigenvalue of ,$\mathbf{T}(\cdot)$ in J, and (13) holds.

- If there exists the kth and lth eigenvalue λ_k and λ_l in $J(k < l)$, then J contains the jth eigenvalue $\lambda_j (k \leq j \leq l)$ as well, and $\lambda_k \leq \lambda_j \leq \lambda_l$.
- Let $\lambda_1 = inf_{x \in D} p(x) \in J$ and $\lambda_l \in J$. If the minimum in (13) is attained for an l dimensional subspace V, then $V \supset D \cup \{0\}$, and (3) can be replaced with

$$\lambda_l = \min_{V \in H_l, V \supset D \cup \{0\}} \sup_{v \in V, v \neq 0} p(v).$$

- $\overline{\lambda}$ is an lth eigenvalue if and only if $\mu = 0$ is the lth eigenvalue of the linear eigenproblem $\mathbf{T}(\lambda)\mathbf{x} = \mu\mathbf{x}$.
- The minimum in (3) is attained for the invariant subspace of $\mathbf{T}(\lambda_l)$ corresponding to its lth largest eigenvalues.

Sylvester's law of inertia has an important role in the nonlinear eigenvalue problems. We will briefly look back to the Sylvester's law of inertia. With this purpose we define the inertia of the Hermitian matrix $\mathbf{T}$ as follows [9].

Definition 3.1.1. The inertia of a Hermitian matrix $\mathbf{T}$ is the triplet of nonnegative integers $\text{In}(\mathbf{T}) = (n_p, n_n, n_z)$ where n_p, n_n and n_z are the number of positive, negative, and zero eigenvalues of $\mathbf{T}$ (counting multiplicities).

Next, we consider a case that an extreme eigenvalue $\lambda_1 := inf_{x \in D} p(x)or\lambda_n := sup_{x \in D} p(x)$ is contained in J.

Theorem 3.1.2 Assume that $T : J \to \mathbb{C}^{n\times n}$ satisfies the conditions of the minmax characterization, and let (n_p, n_n, n_z) be the inertia of $\mathbf{T}(\sigma)$ for some $\sigma \in J$.

- If $\lambda_1 := inf_{x\in D} p(x) \in J$, then the nonlinear eigenproblem $\mathbf{T}(\lambda)\mathbf{x} = \mathbf{0}$ has exactly n_p eigenvalues $\lambda_{n-n_n+1} \le \cdots \le \lambda_n \text{in } J$ in J which are less than σ.
- If $sup_{x\in D} p(x) \in J$, then the nonlinear eigenproblem $\mathbf{T}(\lambda)\mathbf{x} = \mathbf{0}$ has exactly n_n eigenvalues $\lambda_{n-n_n+1} \le \cdots \le \lambda_n \text{in } J$ exceeding σ.

We consider the quadratic eigenvalue problem in the label QEP .That is why we adapt to the real scalar equation (11) QEPu. In this way, we get

$$f(\lambda;x;:=)\lambda^2 x^H Ax + \lambda x^H Bx + x^H Cx = 0 \quad \text{for every fixed } x \in \mathbb{C}^n,\ x \ne 0$$

Natural candidates for the Rayleigh functionals of QEPa (**Eq. 10**) are

$$p_+(\mathbf{x}) := -\frac{\mathbf{x}^H \boldsymbol{B}\mathbf{x}}{2\mathbf{x}^H \boldsymbol{A}\mathbf{x}} + \sqrt{\left(\frac{\mathbf{x}^H \boldsymbol{B}\mathbf{x}}{2\mathbf{x}^H \boldsymbol{A}\mathbf{x}}\right)^2 - \frac{\mathbf{x}^H C\mathbf{x}}{\mathbf{x}^H \boldsymbol{A}\mathbf{x}}} \quad \text{and}$$

$$p_-(\mathbf{x}) := -\frac{\mathbf{x}^H \boldsymbol{B}\mathbf{x}}{2\mathbf{x}^H \boldsymbol{A}\mathbf{x}} - \sqrt{\left(\frac{\mathbf{x}^H \boldsymbol{B}\mathbf{x}}{2\mathbf{x}^H \boldsymbol{A}\mathbf{x}}\right)^2 - \frac{\mathbf{x}^H C\mathbf{x}}{\mathbf{x}^H \boldsymbol{A}\mathbf{x}}}$$

The Rayleigh functionals are the generalization of the Rayleigh quotient.

In this section we deal with the hyperbolic quadratic pencil as an example overdamped problems and gyroscopically stabilized pencil as an example of the changes that are not.

Now, let us look briefly the hyperbolic quadratic pencil. It is overdamped square pencil given by (10) in which the $\mathbf{A} = \mathbf{A}^H > 0$, $\mathbf{B} = \mathbf{B}^H$, $\mathbf{C} = \mathbf{C}^H$. For hyperbolic quadratic pencil, the following interesting features

The ranges $J_+ := p_+(\mathbb{C}^n \backslash \{\mathbf{0}\})$ and $J_- := p_-(\mathbb{C}^n \backslash \{\mathbf{0}\})$ are disjoint real intervals with max $J_- <$ min J_+. $\mathbf{Q}(\lambda)$ is the positive definite for $\lambda <$ min J_- and $\lambda >$ max J_+, and it is the negative definite for $\lambda \in (\max J_-, \min J_+)$.

(Q, J_+) and $(-Q, J_-)$ satisfy the conditions of the variational characterization of the eigenvalues, i.e. there exist $2n$ eigenvalues[1].

$$\lambda_1 \le \lambda_2 \le \cdots \le \lambda_n < \lambda_{n+1} \le \cdots \le \lambda_{2n}$$

and

$$\lambda_j = \min_{\dim V = j} \max_{x \in V,\ x \ne 0} p_-(\mathbf{x}), \quad \lambda_{n+j} = \min_{\dim V = j} \max_{x \in V, x \ne 0} p_+(\mathbf{x}), \quad j = 1, 2, \cdots, n.$$

Now we will look at gyroscopically stabilized system in the label GSS. A quadratic polynomial matrix

$$Q(\lambda)=\lambda^2 \boldsymbol{I}+\lambda \boldsymbol{B}+C,\ \ \boldsymbol{B}=\boldsymbol{B}^H,\ \det(\boldsymbol{B})\neq 0,\ \boldsymbol{C}=\boldsymbol{C}^H$$

is gyroscopically stabilized if for some $k>0$ it holds that

$$|\boldsymbol{B}|>k\boldsymbol{I}+k^{-1}\boldsymbol{C},$$

where denotes the positive square root of $\mathbf{B}^2$.

Definition 3.1.2. A eigenvalue λ is positive type if applies $x^H Q'(\lambda)x>0\quad x\in\mathbb{C}^n,\quad x\neq 0$

A eigenvalue λ is positive type if applies $x^H Q'(\lambda)x<0\quad x\in\mathbb{C}^n,\quad x\neq 0$

Theorem (Barkwell, Lancaster, Markus 1992)

- The spectrum of a gyroscopic stabilized pencil is real, i.e. Q is quasi-hyperbolic.
- All eigenvalues are either of positive type or of negative type
- If (n_p, n_n, n_z) is the inertia of B, then $\mathbf{Q}(\lambda)\mathrm{x} = \mathbf{0}$ has $2n_p$ negative and $2\ n_n$ positive eigenvalues.
- The $2n_p$ negative eigenvalues lie in two disjoint intervals, eigenvalues in each; the ones in the left interval are of negative type, the ones in the right interval are of positive type.
- The $2\ n_n$ negative eigenvalues lie in two disjoint intervals, eigenvalues in each; the ones in the left interval are of negative type, the ones in the right interval are of positive type.

Without loss of generality we will observe only positive eigenvalues value

Let now $p_+(\mathbf{x}):=-\frac{\mathbf{x}^H B\mathbf{x}}{2\mathbf{x}^H\mathbf{x}}+\sqrt{\left(\frac{\mathbf{x}^H B\mathbf{x}}{2\mathbf{x}^H\mathbf{x}}\right)^2-\frac{\mathbf{x}^H C\mathbf{x}}{\mathbf{x}^H\mathbf{x}}}$ and $p_-(\mathbf{x}):=-\frac{\mathbf{x}^H B\mathbf{x}}{2\mathbf{x}^H\mathbf{x}}-\sqrt{\left(\frac{\mathbf{x}^H B\mathbf{x}}{2\mathbf{x}^H\mathbf{x}}\right)^2-\frac{\mathbf{x}^H C\mathbf{x}}{\mathbf{x}^H\mathbf{x}}}$ functionals appropriate for GSS. With them, we can define the Rayleigh functionals $p_-^+:=\begin{cases} p_-(\mathbf{x}) & \text{za}\quad p_-(\mathbf{x})>0 \\ 0 \text{ else} & \end{cases}$ i

$$p_+^+:=\begin{cases} p_+(\mathbf{x}) & \text{za}\quad p_-(\mathbf{x})>0 \\ 0\ else & \end{cases}$$

Voss and Kostić are defined for this function given interval in which the eigenvalues can minmax characterize.

In order to minmax characterized all eigenvalues, we will introduce new Rayleigh functional. It is a new strategy. With this aim we matrices **B** and **C** as well as vector **x** present in the following format :

$$\boldsymbol{B}=\begin{pmatrix} \boldsymbol{B}_1 & 0 \\ 0 & -\boldsymbol{B}_2 \end{pmatrix}, \boldsymbol{B}_i>0,(i=1,2)$$

$$C=\begin{pmatrix} C_{11} & C_{12} \\ C_{12}^H & C_{22} \end{pmatrix}, C_{ii}>0, (i=1,2)$$

$$\mathbf{x}=\begin{pmatrix} z \\ y \end{pmatrix} \neq 0.$$

We define

$$Q_1(\lambda):=\lambda^2 I+\lambda B_1+C_{11}$$
$$Q_2(\lambda):=\lambda^2 I-\lambda B_2+C_{22}$$
$$T(\lambda):=Q_2(\lambda)-C_{12}^H\left(Q_1(\lambda)\right)^{-1}C_{12}$$

Because of the conditions (20) are $\mathbf{Q}_1(\lambda)$ **I** $\mathbf{Q}_2(\lambda)$ are hyperbolic.

We will observe the following problem inherent value

$$T(\lambda)y=0, y\neq 0, \mathbf{y}\in \mathbb{C}^{n-n_p}, \mathbf{y}\neq \mathbf{0}$$

Applies following theorem

Theorem 3.1.3 λ is eigenvalue $\mathbf{Q}(\cdot)$ if and only if λ is eigenvalue $\mathbf{T}(\cdot)$

Proof

$$Q(\lambda)\mathbf{x}=0 \Leftrightarrow \begin{pmatrix} Q_1(\lambda) & C_{12} \\ C_{12}^H & Q_2(\lambda) \end{pmatrix}\mathbf{x}=0 \Leftrightarrow \begin{pmatrix} Q_1(\lambda) & C_{12} \\ C_{12}^H & Q_2(\lambda) \end{pmatrix}\begin{pmatrix} \mathbf{z} \\ \mathbf{y} \end{pmatrix}=0$$

$$\Leftrightarrow \begin{pmatrix} Q_1(\lambda)z+C_{12}y \\ C_{12}^H\mathbf{z}+Q_2(\lambda)\mathbf{y} \end{pmatrix}=\begin{pmatrix} \mathbf{0} \\ \mathbf{0} \end{pmatrix} \Leftrightarrow \left(Q_1(\lambda)z+C_{12}y=0 \wedge C_{12}^H\mathbf{z}+Q_2(\lambda)\mathbf{y}=\mathbf{0}\right)$$

$$\Leftrightarrow \left(z=-\left(Q_1(\lambda)\right)^{-1}C_{12}y \wedge \left(Q_2(\lambda)-C_{12}^H\left(Q_1(\lambda)\right)^{-1}C_{12}\right)\mathbf{y}=0\right)$$

$$\Leftrightarrow \left(z=-\left(Q_1(\lambda)\right)^{-1}C_{12}y \wedge T(\lambda)\mathbf{y}=0\right).$$

Theorem is proved.

Analogous to the (14) we defined the following functions

$$q(\lambda;\mathbf{y}):=\mathbf{y}^{\mathbf{H}}T(\lambda)\mathbf{y} \quad \text{and} \quad f_2(\lambda;\mathbf{y}):=\mathbf{y}^{\mathbf{H}}Q_2(\lambda)\mathbf{y}$$

In the following theorem we give information about the properties $q(\lambda; \mathbf{y})$

Theorem 3.1.4. Function $q(\lambda; \mathbf{y})$ has the following characteristics

- For each vector $q(0; \mathbf{y}) > 0$
- For λ>0 a function $q(\lambda; \mathbf{y})$ exactly two zeros for each vector y. Minimum zero function $q(\lambda; \mathbf{y})$ and is lower than the minimum zero function $f_2(\lambda; \mathbf{y})$ and the largest zero function $q(\lambda; \mathbf{y})$ is greater than the maximum zero function $f_2(\lambda; \mathbf{y})$

◉ From $f_2'(\lambda;\mathbf{y})>0$ $i\lambda>0$ follow $q'(\lambda;\mathbf{y})$

Proof

◉ Because **C** > 0 we have

$$\begin{pmatrix}-\mathbf{y}^H C_{12}^H (C_{11})^{-1} & \mathbf{y}^H\end{pmatrix}\cdot\begin{pmatrix}C_{11} & C_{12}\\ C_{12}^H & C_{22}\end{pmatrix}\cdot\begin{pmatrix}-(C_{11})^{-1}C_{12}\mathbf{y}\\ \mathbf{y}\end{pmatrix}>0.$$

It follows

$$\begin{pmatrix}-\mathbf{y}^H C_{12}^H (C_{11})^{-1} & \mathbf{y}^H\end{pmatrix}\cdot\begin{pmatrix}\mathbf{0}\\ C_{22}\mathbf{y}-C_{12}^H (C_{11})^{-1}C_{12}\mathbf{y}\end{pmatrix}>0.$$

then

$$q(0;\mathbf{y})=\mathbf{y}^H\left(C_{22}-C_{12}^H (C_{11})^{-1}C_{12}\right)\mathbf{y}>0$$

◉ We have already mentioned that $\mathbf{Q}_2(\lambda)$ is hyperbolic. This means that the function $f_2(\lambda;\mathbf{y})$ for each vector **y** has two different real roots. For λ > 0 ist $\mathbf{Q}_1(\lambda)>0$. Therefore, applies

$$q(\lambda;\mathbf{y})<\mathbf{y}^H Q_2(\lambda)\mathbf{y}-\underbrace{\mathbf{y}^H C_{12}^H (Q_1(\lambda))^{-1}C_{12}\mathbf{y}}_{>0}$$

$$<\mathbf{y}^H Q_2(\lambda)\mathbf{y}=f_2(\lambda;\mathbf{y})$$

Of further

$$\lim_{\lambda\to+\infty} q(\lambda;\mathbf{y})=+\infty$$

Of (a) and because (21) and (22) follows that the function $q(\lambda;\mathbf{y})$ for λ>0 has two zero. Minimum zero function $q(\lambda;\mathbf{y})$ *a* and is lower than the minimum zero function $f_2(\lambda;\mathbf{y})$ and the largest zero function $q(\lambda;\mathbf{y})$ *b* is greater than the maximum zero function $f_2(\lambda;\mathbf{y})$

◉ $q'(\lambda;y)=2\lambda y^H y-y^H B_2 y+\underbrace{y^H C_{12}^H (Q_1(\lambda))^{-1}C_{12}y}_{>0}$

Theorem is proved.

Define a new functional

Definition 3.1.3. Let $q(a, y) = q(b, y) = 0$ *and* $0 < a < b$. We define two new functional

$$t_{-}(y) := a$$
$$t_{+}(y) := b$$
$$W_{\pm} := t_{\pm}(h)$$

Theorem 3.1.5. Applies *$maxW_{-} < minW_{+}$*

Proof

$$p_{2\pm}(\mathbf{x}) := -\frac{\mathbf{x}^H B_2 \mathbf{x}}{2\mathbf{x}^H \mathbf{x}} \pm \sqrt{\left(\frac{\mathbf{x}^H B_2 \mathbf{x}}{2\mathbf{x}^H \mathbf{x}}\right)^2 - \frac{\mathbf{x}^H C_2 \mathbf{x}}{\mathbf{x}^H \mathbf{x}}}$$

$$J_{\pm} := p_{2\pm}\left(\mathbb{C}^{n-n_p} \setminus \{\mathbf{0}\}\right).$$

$$maxJ_{-} < minJ_{+},$$

Because $\mathbf{Q}_2(\lambda)$ is hyperbolic.

$$t_{-}(y) < maxJ_{-} < minJ_{+} < t_{+}(y), \text{for every } \mathbf{y} \in \mathbb{C}^{n-n_p} \setminus \{\mathbf{0}\}$$

$$\Rightarrow maxW_{-} < maxJ_{-} < minJ_{+} < maxW_{+}$$

Theorem is proved.

Theorem 3.1.6. All the positive eigenvalues of (19) are either the value of the MinMax of $t_{-}(y)$ or Maxmin of $t_{+}(y)$

3.2. Linearization

This section will focus on the process of linearization. Linearization is a commonly used method to simplify the computation of eigenvalues by lowering the Quadratic Eigenvalue Problem (QEP) on Generalized Eigenvalue Problems (GEP). The issue of eigenvalues typically arises while solving differential equations or systems of differential equations. The concept of linearization originated in the field of differential equations, specifically in cases where the order of a second-order differential equation can be reduced by introducing a system of two first-order partial differential equations with two unknown functions.

The basic idea of linearization in QEPa is the introduction of shift $\mathbf{z} = \lambda\mathbf{x}$ u

$$(\lambda^2 A + \lambda B + C)x = 0.$$

Then we get

- $\lambda Az + Bz + Cx = 0$ or
- $\lambda Az + \lambda Bx + Cx = 0.$

The resulting equations are GEP because they can be written respectively in the form of

- $\begin{pmatrix} -B & -C \\ I & O \end{pmatrix}\begin{pmatrix} \mathbf{z} \\ \mathbf{x} \end{pmatrix} = \lambda \begin{pmatrix} A & O \\ O & I \end{pmatrix}\begin{pmatrix} \mathbf{z} \\ \mathbf{x} \end{pmatrix}$

- $\begin{pmatrix} O & C \\ I & O \end{pmatrix}\begin{pmatrix} \mathbf{z} \\ \mathbf{x} \end{pmatrix} = \lambda \begin{pmatrix} A & B \\ O & I \end{pmatrix}\begin{pmatrix} \mathbf{z} \\ \mathbf{x} \end{pmatrix}.$

Since the corresponding GEPs all matrices $2n \times 2n$ to GEP has 2n eigenvalue and therefore appropriate QEP has also 2n eigenvalue. From the above it is clear that linearization is not unambiguous. However, in the choice of linearization QEP is an important factor to maintain symmetry and some spectral properties QEP if it is possible. The application of linearization will be will in the next chapter.

3.3. Physical Background

We look now at the application of eigenvalues of quadratic problems in engineering. The largest review of applications QEP is in the [10].We have already mentioned in the introduction to eigenvalue problem arises in connection with differential equations or systems of differential equations. In structural mechanics the most commonly are used differential equations and therefore the problem of eigenvalues. Note that the ultimate goal is to determine the effect of vibrations on the performance and reliability of the system, and to control these effects.

We will now demonstrate the linearization of QEP on a concrete example from the engineering. Low vibration system on n unknowns are described by the following system of differential equations

$$\boldsymbol{M}\ddot{\mathbf{y}} + \boldsymbol{C}\dot{\mathbf{y}} + \boldsymbol{K}\mathbf{y} = \mathbf{0},$$

where **M** is mass matrix, **C** is viscous damping matrix, and **K** is the stiffness matrix thus Because of the conditions in physics **M** and **K** are related to the kinetic and strain energy, respectively, by a quadratic form which makes them symmetric. For most structures **M** and **K** are positive definite and space.

The introduction of shift $\mathbf{y} = \mathbf{x}e^{\lambda}$ after rearrangement, we get

$$\left(\lambda^2 \boldsymbol{M} + \lambda \boldsymbol{C} + \boldsymbol{K}\right)\mathbf{x}e^{\lambda} = 0$$

Respectively

$$\left(\lambda^2 \boldsymbol{M} + \lambda \boldsymbol{C} + \boldsymbol{K}\right)\mathbf{x} = 0$$

Therefore, the system (23) has a nontrivial solution **y** is selected, such that λ that QEP (24) has a nontrivial solution **x**.

Now we're going to QED (24) apply linearization method presented in section 3.2. Thus we have the appropriate GEP

$$\begin{pmatrix} -C & -K \\ I & O \end{pmatrix}\begin{pmatrix} \mathbf{z} \\ \mathbf{x} \end{pmatrix} = \lambda \begin{pmatrix} M & O \\ O & I \end{pmatrix}\begin{pmatrix} \mathbf{z} \\ \mathbf{x} \end{pmatrix}$$

When the system is undamped **(C=O)** we get

$$\omega \boldsymbol{M}\mathbf{x} := \lambda^2 \boldsymbol{M}\mathbf{x} = \boldsymbol{K}\mathbf{x} = 0.$$

Because the most common matrix **M** and **K** are symmetric, positive definite and space obtained GEP is easy to solve.

4. CONCISE SUMMARY

Because of the great practical application eigenvalue problem occupies an important place in linear algebra. In this chapter, we discussed the linear and quadratic eigenvalues. In particular, an emphasis is on numerical methods such as the QR algorithm, Rayleigh quotient iteration for linear problems of eigenvalues and linearization and minmax characterization of quadratic problems eigenvalues. The whole chapter shows that the structure of the matrix, participating in the problem of eigenvalues, strongly influence the choice of the method itself. It is also clear that using the features of the structure matrix can do much more effectively existing algorithms. Thus, further studies are going to increase of use feature matrix involved in the problem of eigenvalues, with the aim of improving the effectiveness of the method. Finally, we point out that in this chapter we introduced new Rayleigh functionals for gyroscopically stabilized system that enables complete minmax (maxmin) characterization of eigenvalues. It's a new strategy. We have proved all relevant features of new Rayleigh functionals.

REFERENCES

- Kostić, A.Methods for the determination of some extremal eigenvalues of a symmetric Toeplitz matrix [thesis]. Hamburg: Germany TUHH
- Parllet, B. N.: The Symmetric Eigenvalue Problem. SIAM Classicis in Appied Mathematihs 20, Philadelphia, 1998. DOI: http://dx.do.org/20.1137/1.9781611971163.fm1998,
- Cuppen, J. M.: A divide and conquer method for the symmetric tridiagonal eigenproblem. Numer. Math. 1981; 36: 177–195. DOI: 10.1007/BFo1396757
- Gu, M & Eisenstat A.: Divide-and-conquer algorithm for the symmetric tridiagonal eigenproblem. SIAM J. Matr. Anal. Appl. 1995; 16: 172–191. DOI:10.1137/S0895479892241287

- Chapra, S. T & Canale R. P.: Numerical methods for Engineers. McGraw-Hill Book Company, Singapore, 1990. DOI: 456789 BJE95432
- Duffin, R. J.: A minimax theory for overdamped networks, J. Rat. Mech. Anal., 1955; 4: 221–233.
- Rogers, E.: A minimax theory for overdamped systems, Arch. Ration. Mech. Anal., 1964; 16: 89–96. DOI: 10.1007/BF00281333
- Voss, H.: A minmax principle for nonlinear eigenproblems depending continuously on the eigenparameter, Numer. Lin. Algebra Appl. 2009; 16: 899–913. DOI: 10.1002/nla.670
- Kostić, A & Voss, H.: On Sylvester's law of inertia for nonlinear eigenvalue problems, Electr. Trans. Numer. Anal., 2013; 40: 82–93.
- Tisseur, F & Meerbergen, K.: The quadratic eigenvalue problem, SIAM Review, 2001; 43:235–286.

04 THE ISSUE OF NONNEGATIVE INVERSE ELEMENTARY DIVISORS

1. INTRODUCTION: AN OVERVIEW

Inverse problems are widespread in multiple disciplines and can appear in numerous manifestations. Inverse eigenvalue problems are a notable subset of inverse problems that arise in the realm of mathematical modeling and parameter identification. An illustration of how these challenges might be utilized is in the development of Leontief models in the realm of economics. Inverse eigenvalue problems entail the generation of a specific matrix using provided spectral data. Any inverse eigenvalue problem is characterized by two essential components: establishing the existence of a solution and constructing a solution matrix. The arrangement of the solution matrix, which is usually not singular, is crucial in the analysis of inverse eigenvalue problems. It is crucial to precisely express the problem, as not doing so could cause it to lose significance or importance. Chu and Golub [1] say that "an inverse eigenvalue problem should always be a structured problem." In this chapter, we study the *Nonnegative Inverse Elementary Divisors Problem* (hereafter, the *NIEDP*), which is the problem of finding necessary and sufficient conditions for the existence of a nonnegative matrix with prescribed elementary divisors.

Let A be an $n \times n$ complex matrix, and let

$$J(A) = S^{-1}AS = \begin{bmatrix} J_{n_1(\lambda_1)} & 0 & \ddots & 0 \\ 0 & J_{n_2(\lambda_2)} & \ddots & \ddots \\ \ddots & \ddots & \ddots & 0 \\ 0 & \ddots & 0 & J_{n_k(\lambda_k)} \end{bmatrix}$$

be its *Jordan canonical form* (hereafter, *JCF*). The $n_i \times n_i$ submatrices

$$J_{n_i}(\lambda_i) = \begin{bmatrix} \lambda_i & 1 & & \\ & \lambda_i & \ddots & \\ & & \ddots & 1 \\ & & & \lambda_i \end{bmatrix}, \quad i = 1, 2, \ldots, k$$

are called the *Jordan blocks* of $J(A)$. The *elementary divisors* of A are the polynomials $(\lambda-\lambda_i)^{n_i}$, that is, the characteristic polynomials of $J_{n_i}(\lambda_i)$, $i = 1$,

..., k. The inverse elementary divisors problem (*IEDP*) is the problem of determining necessary and sufficient conditions under which the polynomials $(\lambda-\lambda_1)^{n_1}, (\lambda-\lambda_2)^{n_2}, \ldots, (\lambda-\lambda_k)^{n_k}, n_1+\cdots+n_k=n$, are the elementary divisors of an $n \times n$ matrix A. It is clear that for any arbitrarily prescribed Jordan canonical form J, and for any nonsingular matrix S, there exists a matrix $A = SJS^{-1}$ with J as its *JCF*. In order that the problem be meaningful, the matrix A is required to have a particular structure. When A is required to be an entrywise nonnegative matrix, the problem is called the *nonnegative inverse elementary divisors problem* (*NIEDP*) (see [2–4]). The *NIEDP* is strongly related to another inverse problem, the *nonnegative inverse eigenvalue problem* (hereafter, the *NIEP*), which is the problem of determining necessary and sufficient conditions for a list of complex numbers $\Lambda = \{\lambda_1, \lambda_2, \ldots, \lambda_n\}$ to be the spectrum of an $n \times n$ entrywise nonnegative matrix. If there exists a nonnegative matrix A with spectrum Λ, we say that Λ is realizable and that A is the realizing matrix. The *NIEDP* contains the *NIEP*, and both problems are equivalent if the prescribed eigenvalues are all distinct. Both problems remain unsolved (the *NIEP* is solved only for $n \leq 4$). A number of sufficient conditions or realizability criteria for the *NIEP* to have a solution are known in the literature regarding the problem (see [5] and the references therein). In contrast, only a few works are known about the *NIEDP* (see [3, 4, 6–12]) According to Minc [2], the *NIEDP* looks for to give an answer to the question: which matrices are similar to a nonnegative matrix?

A matrix $A=(a_{ij})_{i,j=1}^{n}$ is said to have *constant row sums* if all its rows sum up to the same constant, say α, i.e.

$$\sum_{j=1}^{n} a_{ij} = \alpha, \ i = 1,\ldots,n.$$

The set of all matrices with constant row sums equal to α is denoted by CS_α. It is clear that any matrix in CS_α has the eigenvector $\mathbf{e} = (1, 1, \ldots 1)^T$ corresponding to the eigenvalue α. A nonnegative matrix A is called stochastic if $A \in CS_1$ and it is called doubly stochastic if $A, A^T \in CS_1$. For lack of simplicity, we shall call *nonnegative generalized stochastic* to a nonnegative matrix $A \in CS_\alpha$ and *nonnegative generalized doubly stochastic* to a nonnegative matrix A with $A, A^T \in CS_\alpha$. The relevance of matrices with constant row sums is due to the well-known fact that if $\Lambda = \{\lambda_1, \lambda_2, \ldots, \lambda_n\}$ is the spectrum of a nonnegative matrix, then Λ is also the spectrum of a nonnegative matrix with constant row sums equal to its Perron eigenvalue (spectral radius).

Denote by $\mathbf{e}_k$ the vector with one in the k-th position and zeros elsewhere. Let S be a nonsingular matrix such $S^{-1}AS = J(A)$ is the *JCF* of A. If $A \in CS_{\lambda_1}$, S can be chosen so that $S\mathbf{e}_1 = \mathbf{e}$ and, in this case, it is easy to see that the rows of $S^{-1} = (\hat{s}_{ij})$ satisfy:

$$\sum_{j=1}^{n} \hat{s}_{1j} = 1 \text{ and } \sum_{j=1}^{n} \hat{s}_{ij} = 0, i = 2,\ldots,n.$$

If T is an $n \times n$ matrix of the form

$$T = \begin{bmatrix} \lambda_1 & * & \cdots & * \\ 0 & * & \ddots & \vdots \\ \vdots & \vdots & \ddots & * \\ 0 & * & \cdots & * \end{bmatrix}, \text{ and } S = \begin{bmatrix} 1 & s_{12} & \cdots & s_{1n} \\ 1 & s_{22} & \ddots & s_{2n} \\ \vdots & \ddots & \ddots & \ddots \\ 1 & s_{n2} & \ddots & s_{nn} \end{bmatrix}$$

is nonsingular, then $STS^{-1}\mathbf{e} = \lambda_1\mathbf{e}$, that is, $STS^{-1} \in CS_{\lambda 1}$. We shall denote by E_{ij} the $n \times n$ matrix with 1 in the $(i, j)^{th}$ position and zeros elsewhere. The following simple perturbation allows us to join two or more Jordan blocks corresponding to a same eigenvalue λ_p to obtain one Jordan block of a bigger size: Let A be an $n \times n$ matrix with$JCFJ(A)$.LettheJordanblocks $J_{m_1}(\lambda_p), J_{m_2}(\lambda_p), \ldots, J_{m_q}(\lambda_p)$ becorrespondingtothe eigenvalue λ_p. Let $\xi = \sum_{j=1}^{p-1} m_j$, $1 \le p \le k$, with $\xi = 0$ if $p = 1$, and let $E = \sum_{i \in K} E_{i,i+1}$. Then by using the perturbation $J(A) + E$, with

$$K = \left\{ \xi + m_1, \xi + m_1 + m_2, \ldots, \xi + \sum_{i=1}^{q-1} m_i \right\},$$

we obtain a Jordan block of bigger size $J_\gamma(\lambda_p)$, corresponding to the elementary divisor $(\lambda - \lambda_p)^\gamma$, $\gamma = m_1 + m_2 + \ldots + m_q$.

Observe that if $E = \sum_{i \in K} E_{i,i+1}$ with $K \subset \{1, 2, \ldots, n-1\}$, and A is an $n \times n$ complex matrix with JCF $J(A) = S^{-1}AS$, then for an appropriate set K

$$J(A) + E = S^{-1}AS + E = S^{-1}\left(A + SES^{-1}\right)S$$

is the JCF of $A + SES^{-1}$. If $A \in CS_{\lambda 1}$ and $S = [\mathbf{e}| * | \cdots |*]$, then

$$\left(A + SES^{-1}\right) \in cs_{\lambda_1}.$$

The first works on the *NIEDP* are due to H. Minc [3, 4]. Minc studied the problem for nonnegative and doubly stochastic matrices, modulo the *NIEP*. In particular, he proved the following two results, which we collect as:

Theorem 1 Minc [3] *Let $\Lambda = \{\lambda_1, \lambda_2, \ldots, \lambda_n\}$ be a list of complex numbers, which is realizable by a diagonalizable positive (diagonalizable positive doubly stochastic) matrix A. Then, for each JCF J_Λ associated with Λ, there exists a positive (positive doubly stochastic) matrix B with the same spectrum as A, and with JCF $J(B) = J_\Lambda$.*

According to Minc, the positivity condition is essential in his proof, and it is not known if the result holds without this condition (see [2]). Specifically, it is not known: *i)* whether for every positive matrix, there exists a diagonalizable positive matrix with the same spectrum, *ii)* whether for every nonnegative diagonalizable matrix with spectrum $\Lambda = \{\lambda_1, \ldots, \lambda_n\}$, there exists a nonnegative matrix for each *JCF* associated with Λ.

Usually, to work with the *NIEDP*, we are given a list of complex numbers $\Lambda = \{\lambda_1, \lambda_2, ..., ;_n\}$, from which we want to construct a nonnegative or positive matrix with spectrum Λ and with prescribed elementary divisors. In this sense, we mention two matrix perturbation results, which have been employed in connection with the *NIEP* and the *NIEDP*, to derive sufficient conditions for the existence and construction of nonnegative matrices with prescribed spectrum and prescribed elementary divisors. The first result, due to Brauer [13, Theorem 27], shows how to change a single eigenvalue of an $n \times n$ matrix, via a rank-1 perturbation, without changing any of the remaining $n - 1$ eigenvalues. The second result, due to R. Rado and introduced by Perfect in [14], is an extension of the Brauer result. It shows how to change r eigenvalues of an $n \times n$ matrix A via a rank-r perturbation, without changing any of the remaining $n - r$ eigenvalues (see [15] to understand how Rado's result is applied to the NIEP). The proof of the Brauer result, which we give here, is due to R. Reams [16].

Theorem 2 Brauer [13] *Let A be an $n \times n$ arbitrary matrix with eigenvalues $\lambda_1, ..., \lambda_n$. Let $\mathbf{v} = (v_1,..., v_n)^T$ an eigenvector of A associated with the eigenvalue λ_k and let$\mathbf{q} = (q_1,..., q_n)^T$ be any n-dimensional vector. Then the matrix $A + \mathbf{vq}^T$ has eigenvalues $\lambda_1, ..., \lambda_{k-1}, \lambda_k + v^T q, \lambda_{k+1}, ..., \lambda_n$.*

Proof. [16] From the Schur's triangularization theorem, let U be an $n \times n$ nonsingular matrix such that

$$U^{-1}AU = \begin{bmatrix} \lambda_1 & * & \cdots & * \\ & \lambda_2 & \ddots & \vdots \\ & & \ddots & * \\ & & & \lambda_n \end{bmatrix}$$

is an upper triangular matrix, with $\mathbf{v}$ being the first column of U. Then,

$$U^{-1}\left(A + \mathbf{vq}^T\right)U = U^{-1}AU + \begin{bmatrix} q_1 & q_2 & \cdots & q_n \\ & & & \\ & & & \\ & & & \end{bmatrix} U$$

$$= \begin{bmatrix} \lambda_1 + \mathbf{q}^T\mathbf{v} & * & \cdots & * \\ & \lambda_2 & \ddots & \vdots \\ & & \ddots & * \\ & & & \lambda_n \end{bmatrix}.$$

and the result follows.■

Theorem 3 Rado [14] *Let A be an $n \times n$ arbitrary matrix with eigenvalues $\lambda_1, ..., \lambda_n$ and let $\Omega = diag\{\lambda_1, ..., \lambda_r\}$ for some $r \leq n$. Let X be an $n \times r$ matrix with rank r such that its columns $x_1, x_2, ..., x_r$ satisfy $Ax_i = \lambda_i x_i$, $i = 1, ..., r$. Let C be an $r \times n$ arbitrary matrix. Then the matrix $A + XC$ has eigenvalues $\mu_1, ..., \mu_r, \lambda_{r+1}, ..., \lambda_n$, where $\mu_1, ..., \mu_r$ are eigenvalues of the matrix $\Omega + CX$.*

Proof. Let $S = [X|Y]$ a nonsingular matrix with $S^{-1} = \left[\begin{smallmatrix} U \\ V \end{smallmatrix}\right]$. Then $UX = I_r$, $VY = I_{n-r}$, and $VX = 0$, $UY = 0$. Let $C = [C_1|C_2]$, $X = \left[\begin{smallmatrix} X_1 \\ X_2 \end{smallmatrix}\right]$, $Y = \left[\begin{smallmatrix} Y_1 \\ Y_2 \end{smallmatrix}\right]$. Then, since $AX = X\Omega$,

$$S^{-1}AS = \begin{bmatrix} U \\ V \end{bmatrix} [X\Omega \mid AY] = \begin{bmatrix} \Omega & UAY \\ 0 & VAY \end{bmatrix}$$

and

$$S^{-1}XCS = \begin{bmatrix} I_r \\ 0 \end{bmatrix} [C_1 \mid C_2] S = \begin{bmatrix} C_1 & C_2 \\ 0 & 0 \end{bmatrix} \begin{bmatrix} X_1 & Y_1 \\ X_2 & Y_2 \end{bmatrix} = \begin{bmatrix} CX & CY \\ 0 & 0 \end{bmatrix}.$$

Thus,

$$S^{-1}(A + XC)S = S^{-1}AS + S^{-1}XCS = \begin{bmatrix} \Omega + CX & UAY + CY \\ 0 & VAY \end{bmatrix},$$

and we have $\sigma(A + XC) = \sigma(\Omega + CX) + \sigma(A) - \sigma(\Omega)$.■

The following result in [6], which will be frequently used later, shows how is the *JCF* of the Brauer perturbation $A + \mathbf{eq}^T$.

Lemma 1 [6] *Let* $A \in CS_{\lambda 1}$ *be with JCF*

$$J(A) = S^{-1}AS \ldots, J_{n_k}(\lambda_k)\}$$

$$\lambda_1 + \sum q_i \neq \lambda_i, i = 2, \ldots, n.$$

$J(A) + \left(\sum_{i=1}^{n} q_i\right) E_{11}$ $(q_1, \ldots, q_n)$ *ar* . *Then the JCF of* $A + \mathbf{eq}^T$ *is* . *In particular,* $if \sum_{i=1}^{n} q_i = 0$, *then A and* $A + \mathbf{eq}^T$ *are similar.*

2. CASES COMPLETELY SOLVED

In this section, we shall discuss about certain subproblems of the *NIEDP*, for which a complete solution has been obtained. We start with a list of real nonnegative numbers $\Lambda = \{\lambda_1, \lambda_2, \ldots, \lambda_n\}$ with $\lambda_1 > \lambda_2 \geq \cdots \geq \lambda_n \geq 0$. Let $D = diag\{\lambda_1, \lambda_2, \ldots, \lambda_n\}$. In [17], Perfect shows that the matrix

$$P = \begin{bmatrix} 1 & 1 & \cdots & 1 & 1 \\ 1 & 1 & \cdots & 1 & -1 \\ \vdots & \vdots & \cdots & -1 & 0 \\ 1 & 1 & \ddots & \ddots & \vdots \\ 1 & -1 & 0 & \cdots & 0 \end{bmatrix}$$

has the property that $PDP^{-1} \in CS_{\lambda 1}$ is positive diagonalizable with spectrum Λ. Then we have:

Theorem 4 [6] *Let* $\Lambda = \{\lambda_1, \ldots, \lambda_n\}$ *be a list of real numbers with* $\lambda_1 > \lambda_2 \geq \cdots \geq \lambda_n \geq 0$. *Then, there exists a nonnegative matrix* $A \in CS_{\lambda 1}$ *with spectrum* Λ, *for each possible JCF associated with* Λ.

Proof. Let $D = diag\{\lambda_1, \lambda_2, \ldots, \lambda_n\}$ and let P be the matrix in (2). Then $PDP-1 \in CS_{\lambda 1}$ is positive (generalized stochastic) with spectrum Λ and linear elementary divisors. Let

$$K \subset \{2,3,\ldots,n-1\} \text{ and } E = \sum_{i \in K} E_{i,i+1},$$

such that $D + E$ is the desired *JCF*. Then there exists $\varepsilon > 0$ such that $A = PDP^{-1} + \varepsilon PEP^{-1}$ is nonnegative, and since $D + \varepsilon E$ and $D + E$ are diagonally similar (with $diag\{1, \varepsilon, \varepsilon^2, \ldots, \varepsilon^{n-1}\}$), A has *JCF* equal to $D + E$. Moreover, since $P\mathbf{e}_1 = \mathbf{e}$, the rows of P^{-1} satisfy (1) and $PEP^{-1}\mathbf{e} = 0$. Thus $A \in CS_{\lambda 1}$ has the desired elementary divisors.■

The following two cases correspond, respectively, to lists $\Lambda = \{\lambda_1, \ldots, \lambda_n\}$ of real numbers satisfying $\lambda_1 > 0 > \lambda_2 \geq \cdots \geq \lambda_n$, which are called lists of Suleimanova type; and to lists $\Lambda = \{\lambda_1, \ldots, \lambda_n\}$ of complex numbers satisfying $\lambda_i \in \mathcal{F}$, $i = 2, \ldots, n$, where

$$F = \{z \in C : \text{Re } z < 0, |\text{Re } z| \geq |\text{Im } z|\},$$

which are called lists of complex Suleimanova type. In both cases, the *NIEP* has a solution if and only if $\sum_{i=1}^{n} \lambda_i \geq 0$ (see [18, 19]).

Let $\Lambda = \{\lambda_1, \ldots, \lambda_n\}$ be a list of complex numbers with $\bar{\Lambda} = \Lambda$, $\sum_{i=1}^{n} \lambda_i \geq 0$, and $\lambda_1 \geq |\lambda_i|$, $i = 2, \ldots, n$. In [9, Theorem 2.1], the authors define

$$M = \max_{2 \leq k \leq n} \{0, Re\lambda_k + Im\lambda_k\}$$

$$m = -\sum_{k=2}^{n} \min\{0, Re\lambda_k, Im\lambda_k\}$$

and show that if $\lambda_1 \geq M + m$, when all possible Jordan blocks $J_{ni}(\lambda_i)$, of size $n_i \geq 2$, are associated to a real eigenvalue $\lambda_i < 0$; or when there is at least one Jordan block $J_{ni}(\lambda_i)$, of size $n_i \geq 2$, associated to a real eigenvalue $\lambda_i \geq 0$ with $M = Re\lambda_{i0} + Im\lambda_{i0}$, for some i_0, then there exists an $n \times n$ nonnegative matrix $A \in CS_{\lambda 1}$ with spectrum Λ and with prescribed elementary divisors. The result in [9, Theorem 2.1] was used by the authors to prove the following

Corollary 1 [9] *Let $\Lambda = \{\lambda_1, \ldots, \lambda_n\}$ be a list of complex numbers with $\lambda_i \in \mathcal{F}$, $i = 2, \ldots, n$. Then, for each possible JCF associated with Λ, there exists a nonnegative matrix $A \in CS_{\lambda 1}$ with spectrum Λ if and only if $\sum_{i=1}^{n} \lambda_i \geq 0$.*

Proof. The condition is necessary for the existence of a nonnegative matrix with spectrum Λ. The condition is also sufficient. In fact, since

$$M = \max_{2 \leq k \leq n} \{0, Re\lambda_k + Im\lambda_k\} = 0,$$

$$m = -\sum_{k=2}^{n} \min\{0, Re\lambda_k, Im\lambda_k\} = -\sum_{k=2}^{n} \text{Re}\lambda_k = -\sum_{k=2}^{n} \lambda_i,$$

then if $\lambda_1 + \lambda_2 + \cdots + \lambda_n \geq 0$, $\lambda_1 \geq -\sum_{k=2}^{n} \lambda_i = M + m$. ■

In [19, Theorem 3.3], the authors show that if $\Lambda = \{\lambda_1, \ldots, \lambda_n\}$ is a list of complex numbers with $\lambda_i \in \mathcal{F}$, $i = 2. \ldots. n$, then there exists a nonnegative matrix with spectrum Λ if and only if $\sum_{i=1}^{n} \lambda_i \geq 0$.. Thus we have:

Corollary 2 [6] *Let $\Lambda = \{\lambda_1, \ldots, \lambda_n\}$ be a list of real numbers with $\lambda_1 > 0 > \lambda_2 \geq \cdots \geq \lambda_n$. Then, for each possible JCF associated with Λ, there exists a nonnegative matrix $A \in CS_{\lambda 1}$ with spectrum Λ if and only if $\sum_{i=1}^{n} \lambda_i \geq 0$.*

In [20], Šmigoc extends the result in [19, Theorem 3.3], to the region

$$\mathcal{G} = \{z \in \mathbb{C} : \operatorname{Re} z < 0, |\sqrt{3}\operatorname{Re} z| \geq |\operatorname{Im} z|\},$$

that is, she shows that if $\Lambda = \{\lambda_1, \ldots, \lambda_n\}$ is a list of complex numbers with $\lambda_i \in G$, $i = 2, \ldots, n$, then there exists a nonnegative matrix with spectrum Λ if and only if

$\sum_{i=1}^{n} \lambda_i \geq 0$.

The following result extends Corollary 1 to lists of complex numbers $\Lambda = \{\lambda_1, \ldots, \lambda_n\}$, with $\lambda_i \in G$, $i = 2, \ldots, n$. Before, to state this result and its proof, we need the following lemmas, which we set here without proof:

Lemma 2 [20] *Let $A = \begin{bmatrix} A_1 & \mathbf{a} \\ \mathbf{b}^T & c \end{bmatrix}$ be an $n \times n$ matrix and let B be an $m \times m$ matrix with JCF $J(B) = \begin{bmatrix} c & 0 \\ 0 & I(B) \end{bmatrix}$. Then the matrix*

$$C = \begin{bmatrix} A_1 & \mathbf{a}\mathbf{t}^T \\ \mathbf{s}\mathbf{b}^T & B \end{bmatrix}, B\mathbf{s} = c\mathbf{s}, \mathbf{t}^T B = c\mathbf{t}^T, \text{with } \mathbf{t}^T\mathbf{s} = 1,$$

has JCF

$$J(C) = \begin{bmatrix} J(A) & 0 \\ 0 & I(B) \end{bmatrix}.$$

Lemma 3 [12] *Let $\Lambda = \{\lambda_1, a \pm bi, \ldots, a \pm bi\}$ be a list of n complex numbers, with $n \geq 3$, $a < 0$, $b > 0$. If*

$$\lambda_1 \geq \max\left\{-(n-1)a, -\frac{(2n-5)a^2 + b^2}{2a}\right\},$$

then for each JCF associated with Λ, there exists an $n \times n$ nonnegative matrix $A = (a_{ij})$, with spectrum Λ and entry $a_{nn} = s_1(\Lambda)$.

Corollary 3 [12] *Let $\Lambda = \{\lambda_1, a \pm bi, \ldots, a \pm bi\}$ be a list of n complex numbers, with $0 < b \leq -\sqrt{3}a$.. Then, for each JCF associated with Λ, there exists a nonnegative matrix $A = (a_{ij})$ with spectrum Λ and entry $a_{nn} = s_1(\Lambda)$, if and only if $\lambda_1 + (n-1)a \geq 0$.*

Proof. It is clear that the condition is necessary. Assume that $\lambda_1 + (n-1)a \geq 0$. Since $0 < b \leq -\sqrt{3}a$,, it follows that

$$\max\left\{-(n-1)a,-\frac{(2n-5)a^2+b^2}{2a}\right\}$$

$$=\max\left\{-(n-1)a,-(n-1)a+\frac{3a^2-b^2}{2a}\right\}$$

$$=-(n-1)a$$

$$\leq \lambda_1.$$

Hence, from Lemma 3, Λ is the spectrum of a nonnegative matrix $A = (a_{ij})$ with entry $a_{nn} = s_1(\Lambda)$, for each *JCF* associated with Λ.

Now we state the main result of this section:

Theorem 5 [12] *Let $\Lambda = \{\lambda_1, \ldots, \lambda_n\}$ be a list of complex numbers with $\lambda_i \in G$, $i = 2, \ldots, n$. Then, for each JCF associated with Λ, there exists a nonnegative matrix A with spectrum Λ and entry $a_{nn}=\sum_{i=1}^{n}\lambda_i$ if and only if $\sum_{i=1}^{n}\lambda_i\geq 0$.*

Proof. It is clear that the condition is necessary. Let $\lambda_1 > 0 > \lambda_2 \geq \cdots \geq \lambda_p$ be $s_1(\Lambda)=\sum_{i=1}^{n}\lambda_i\geq 0$. s and let $\lambda_{p+1}, \lambda_{p+2}=\bar{\lambda}_{p+1}, \ldots, \lambda_{n-1}, \lambda_n=\bar{\lambda}_{n-1}$ be complex nonreal numbers, . Let *m* be the number of distinct pairs of complex conjugate. Consider the partition

$$\Lambda_0=\{\lambda_1,\lambda_2,\ldots,\lambda_p\}$$

$$\Lambda_r=\{s_1(\Lambda_{r-1}),\lambda_{p+r},\bar{\lambda}_{p+r},\ldots,\lambda_{p+r},\bar{\lambda}_{p+r}\}, r=1,\ldots,m$$

with Λ_r having $n_r + 1$ elements, where n is the number of elements of $\Lambda_r - \{s_1(\Lambda_{r-1})\}$, in such a way that $\sum_{r=1}^{m} n_r = n - p$. The list Λ_0 satisfies Corollary 2, and then, we can compute a matrix $A_0=(a_{ij}^{(0)})$ with spectrum Λ_0, the entry $a_{pp}^{(0)}=s_1(\Lambda_0)$ and with arbitrarily prescribed elementary divisors. Moreover, since $0<\operatorname{Im}\lambda_{p+r}\leq-\sqrt{3}\operatorname{Re}\lambda_{p+r}$, $r = 1, \ldots, m$, there exists, from Corollary 3, a nonnegative matrix $A_r=(a_{ij}^{(r)})$ of size $n_r + 1$, of size $n_r + 1$, with spectrum Λ_r, entry $a_{n_r+1,n_r+1}^{(r)}=s_1(\Lambda_r)$,

for each *JCF* associated with Λ_r. Finally, by employing the Lemma 2 (Šmigoc Lemma) *m* times, we construct a nonnegative matrix *A* with spectrum Λ and with arbitrarily prescribed elementary divisors. ■

Example 1 *Let*

$$\Lambda = \{23, -1, -1, -1, -2 \pm 3i, -2 \pm 3i, -3 \pm 5i, -3 \pm 5i\}$$

be given. We want to construct a 12 × 12 *nonnegative matrix with elementary divisors*

$$(\lambda - 23), (\lambda + 1)^2, (\lambda + 1), (\lambda^2 + 4\lambda + 13)^2$$

$$(\lambda + 3 - 5i), (\lambda + 3 - 5i), (\lambda + 3 + 5i), (\lambda + 3 + 5i).$$

Consider the lists

$$\Lambda_0 = \{23, -1, -1, -1\}$$

$$\Lambda_1 = \{20, -2 \pm 3i, -2 \pm 3i\}$$

$$\Lambda_2 = \{12, -3 \pm 5i, -3 \pm 5i\}.$$

Clearly Λ_0 *is realizable bv a nonnegative matrix* A_0, *with elementary divisors* $(\lambda - 23)$, $(\lambda + 1)^2$, $(\lambda + 1)$, *and entry* $a_{44}^{(0)}$ 20. Λ_1 *and* Λ_2 *are also realizable bv nonneaative matrices* A_1, *with elementary divisors* $(\lambda - 20)$, $(\lambda^2 + 4\lambda + 13)^2$, *and entry* $a_{55}^{(1)} = 12$, *and* A_2, *with linear elementary divisors* $(\lambda - 12)$, $(\lambda + 3 \pm 5i)$, $(\lambda + 3 \pm 5i)$, *and entry* $a_{55}^{(2)} = 0$, *respectively (see Corollary 3). By applying Lemma 2 to the matrices* A_0 *and* A_1 *we obtain an* 8 × 8 *nonnegative matrix* $B_1 = (b_{ij})$ *with spectrum* $\Lambda_0 \cup \Lambda_1 - \{20\}$, *elementary divisors*

$$(\lambda - 23), (\lambda + 1)^2, (\lambda + 1), (\lambda^2 + 4\lambda + 13)^2$$

and entry $b_{88} = 12$. *Next, we employ Lemma 2 again with the matrices* B_1 *and* A_2, *to obtain a nonnegative matrix A with spectrum* Λ *and the prescribed elementary divisors.*

The method that we employ to prove Theorem 5, allow us to compute, under certain conditions, a nonnegative matrix *A* with spectrum $\Lambda = \{\lambda_1, \ldots, \lambda_n\}$, where $\lambda_i \in G$, $i = 2, \ldots, n$, for each *JCF* associated with Λ. However, this procedure does not work for any list of complex numbers in the left half plane. In particular, the list

$$\Lambda = \{6, -1+3i, -1-3i, -1+3i, -1-3i\}$$

satisfies the Laffey and Šmigoc conditions [21], and therefore, it is realizable. However, Λ does not satisfy the condition of Lemma 3. As a consequence, we cannot obtain, from Lemma 3, a nonnegative matrix with spectrum Λ and linear elementary divisors.

3. GENERAL SUFFICIENT CONDITIONS

In this section, we start with a general result, which gives a simple sufficient condition for the existence and construction of a nonnegative matrix with prescribed spectrum and elementary divisors. In particular, the result stated that if the Perron eigenvalue λ_1 is large enough, then the list $\Lambda = \{\lambda_1, \ldots, \lambda_n\}$ is the spectrum of a nonnegative matrix with prescribed elementary divisors. Since the result and its proof are somewhat involved, we start with the following example in order to illustrate the ideas and the constructive procedure followed in the proof:

Example 2 *Let* $\Lambda = \{\lambda_1, 3, 3, -1, -1, -1 \pm 3i, -1 \pm 3i\}$.

Case i): We want to construct a nonnegative matrix A with elementary divisors

$$J_1(\lambda_1), J_1(3), J_1(3), J_2(-1), J_2(-1+3i), J_2(-1-3i).$$

We consider the initial matrix

$$D = \begin{bmatrix} \lambda_1 &&&&&&&& \\ & 3 &&&&&&& \\ && 3 &&&&&& \\ &&& -1 &&&&& \\ &&&& -1 &&&& \\ &&&&& -1 & 3 && \\ &&&&& -3 & -1 && \\ &&&&&&& -1 & 3 \\ &&&&&&& -3 & -1 \end{bmatrix}$$

Let $S = [\mathbf{e}|\mathbf{e}_2| \cdots |\mathbf{e}_9]$ *and* $E = E_{4,5} + E_{7,8}$. *Then*

$$B = S(D + \varepsilon E)S^{-1}$$

$$= \begin{bmatrix} \lambda_1 &&&&&&&& \\ \lambda_1 - 3 & 3 &&&&&&& \\ \lambda_1 - 3 && 3 &&&&&& \\ \lambda_1 + 1 - \varepsilon &&& -1 & \varepsilon &&&& \\ \lambda_1 + 1 &&&& -1 &&&& \\ \lambda_1 - 2 &&&&& -1 & 3 && \\ \lambda_1 + 4 - \varepsilon &&&&& -3 & -1 & \varepsilon & \\ \lambda_1 - 2 &&&&&&& -1 & 3 \\ \lambda_1 + 4 &&&&&&& -3 & -1 \end{bmatrix}.$$

It is clear that $B \in CS_{\lambda 1}$ *has the desired JCF. Now, in order that B becomes nonnegative, we take*

$$q^T = (-10, 0, 0, 1, 1, 3, 1, 3, 1), |\varepsilon| \leq 1,$$

and apply Theorem 2 (Brauer result) to obtain the nonnegative matrix $A = B + \mathbf{e}\mathbf{q}^T$, *where* $\lambda_1 \geq 10 + 3 = 13$. *Moreover, from Lemma 1, A has the prescribed elementary divisors. Observe that the values 3 and 10 represent the cost we must pay to obtain A. These values are defined as*

$$M = \max_{2 \le k \le n} \{0, Re\lambda_k + Im\lambda_k\} \ and$$

$$m = -\sum_{k=2}^{n} \min\{0, Re\lambda_k, Im\lambda_k\}.$$

In this case, $M = 3$ *and* $m = 10$.

Case ii): Now, suppose we want to construct a nonnegative matrix A with elementary divisors

$$J_1(\lambda_1), J_2(3), J_2(-1), J_2(-1+3i), J_2(-1-3i).$$

In this case, we take $E = E_{2,3} + E_{4,5} + E_{7,8}$, *and*

$$B = S(D + \varepsilon E)S^{-1}$$

$$= \begin{bmatrix} \lambda_1 & & & & & & & & \\ \lambda_1 - 3 - \varepsilon & 3 & \varepsilon & & & & & & \\ \lambda_1 - 3 & & 3 & & & & & & \\ \lambda_1 + 1 - \varepsilon & & & -1 & \varepsilon & & & & \\ \lambda_1 + 1 & & & & -1 & & & & \\ \lambda_1 - 2 & & & & & -1 & 3 & & \\ \lambda_1 + 4 - \varepsilon & & & & & -3 & -1 & \varepsilon & \\ \lambda_1 - 2 & & & & & & & -1 & 3 \\ \lambda_1 + 4 & & & & & & & -3 & -1 \end{bmatrix}.$$

By choosing

$$q^T = (-10,0,0,1,1,3,1,3,1), \varepsilon > 0,$$

we obtain $A = B + \mathbf{eq}^T$, *which for* $\lambda_1 > M + N = 13$, *is nonnegative with spectrum* Λ, *and with the prescribed elementary divisors.*

Now we state the mentioned result. A proof of it can be found in [9, Theorem 2.1].

Theorem 6 [9] *Let* $\Lambda = \{\lambda_1, \lambda_2, ..., \lambda_n\}$ *be a list of complex numbers with* $\bar{\Lambda} = \Lambda$, $\sum_{i=1}^{n} \lambda_i \ge 0$, *and* $\lambda_1 \ge |\lambda_i|$, $i = 2, ..., n$. *Let*

$$M = \max_{2 \le i \le n} \{0, \mathrm{Re}\ \lambda_i + \mathrm{Im}\ \lambda_i\}; \quad m = -\sum_{i=2}^{n} \min\{0, \mathrm{Re}\ \lambda_i, \mathrm{Im}\ \lambda_i\},$$

if

$$i) \quad \lambda_1 \ge M + m,$$

when all possible Jordan blocks $J_{n_i}(\lambda_i)$, *of size* $n_i \ge 2$, *are associated to a real eigenvalue* $\lambda_i < 0$; *or when there is at least one Jordan block* $J_{n_i}(\lambda_i)$, *of size* $n_i \ge 2$, *associated to a real eigenvalue* $\lambda_i \ge 0$ *with* $M = \mathrm{Re}\lambda_{i0} + \mathrm{Im}\ \lambda_{i0}$, *for some* $i_0 \le n - 1$.

or if

$$\text{ii)}\lambda_1 > M + m,$$

when at least one Jordan block $J_{n_i}(\lambda_i)$, *of size* $n_i \geq 2$, *is associated to a real eigenvalue* $\lambda_i \geq 0$ *with* $M = \lambda_k \geq 0$,

then there exists an $n \times n$ *nonnegative matrix* $A \in CS_{\lambda 1}$ *with spectrum* Λ *and with prescribed elementary divisors*

$$(\lambda - \lambda_1), (\lambda - \lambda_2)^{n_2}, \ldots, (\lambda - \lambda_k)^{n_k}, \; n_2 + \cdots + n_k = n - 1.$$

Let A, X, C, and Ω be as in Theorem 3, with

$$S^{-1}(A + XC)S = \begin{bmatrix} \Omega + CX & UAY + CY \\ 0 & VAY \end{bmatrix},$$

where $B = \Omega + CX$ is an $r \times r$ matrix with eigenvalues $\mu_1, \ldots, \mu_r$ (the new eigenvalues) and diagonal entries $\lambda_1, \ldots, \lambda_r$ (the former eigenvalues). Then from a result in [2, Chapter *VI*, Lemma 1.2], if $B = \Omega + CX$ and VAY have no common eigenvalues, $A + XC$ is similar to $B \oplus VAY$. Then we have:

Lemma 4 *Let A, X, Y, V, C, and* Ω *be as above. If the matrices* $B = \Omega + CX$ *and VAY have no common eigenvalues, then*

$$J(A+XC)=J(B)\oplus J(VAY).$$

In particular, if $CX = 0$, *A and* $A + XC$ *are similar.*

The following result extends Theorem 6:

Theorem 7 [9] *Let* $\Lambda = \{\lambda_1, \lambda_2, \ldots, \lambda_n\}$ *be a list of complex numbers with* $\Lambda = \bar{\Lambda}, \lambda_1 \geq \max_i |\lambda_i|, i = 2, \ldots, n;$ $\sum_{i=1}^{n} \lambda_i \geq 0$. *Let* $\Lambda = \Lambda_0 \cup \Lambda_1 \cup \cdots \cup \Lambda_{p_0}$ *be a pairwise disjoint partition, with* $\Lambda_k = \{\lambda_{k1}, \lambda_{k2}, \ldots, \lambda_{kp_k}\}$; $\lambda_{11} = \lambda_1$, $k = 1, \ldots, p_0$, *where* Λ_0 *is realizable,* p_0 *is the number of elements of the list* Λ_0 *and some lists* Λ_k, $k = 1, \ldots, p_0$, *can be empty. Let* $\omega_1, \ldots, \omega_{p0}$ *be real numbers satisfying* $0 \leq \omega_k \leq \lambda_1$, $k = 1, \ldots, p_0$. *Suppose that*

i) For each $k = 1, \ldots, p_0$, *there exists a list* $\Gamma_k = \{\omega_k, \lambda_{k1}, \ldots, \lambda_{k\,pk}\}$ *with* $\omega_k \geq M_k + m_k$ *or* $\omega_k > M_k + m_k$, *as in Theorem 6, where*

$$M_k = \max_{1 \leq i \leq p_k} \{0, \text{Re } \lambda_{ki} + \text{Im } \lambda_{ki}\}; \; m_k = -\sum_{i=1}^{p_k} \min\{0, \text{Re } \lambda_{ki}, \text{Im } \lambda_{ki}\},$$

and

ii) there exists a $p_0 \times p_0$ *nonnegative matrix B with spectrum* Λ_0 *and diagonal entries* $\omega_1, \ldots, \omega_{p0}$.

Then there exists an $n \times n$ *nonnegative matrix* $M \in CS_{\lambda 1}$ *with spectrum* Λ *and with prescribed elementary divisors.*

Proof. Let

$$D_k = \begin{bmatrix} \omega_k & & & & & & & \\ & \ddots & & & & & & \\ & & \lambda_{ks} & & & & & \\ & & & \operatorname{Re}\lambda_{k(s+1)} & -\operatorname{Im}\lambda_{k(s+1)} & & & \\ & & & \operatorname{Im}\lambda_{k(s+1)} & \operatorname{Re}\lambda_{k(s+1)} & & & \\ & & & & & \ddots & & \\ & & & & & & \operatorname{Re}\lambda_{k(p_k-1)} & -\operatorname{Im}\lambda_{k(p_k-1)} \\ & & & & & & \operatorname{Im}\lambda_{k(p_k-1)} & \operatorname{Re}\lambda_{k(p_k-1)} \end{bmatrix}$$

with spectrum Γ_k and let $E_\iota = \Sigma E_{i,i+1}$ such that $D_\iota + \varepsilon E_\iota$ is the prescribed *JCF*. Let $S_k = [\mathbf{e} \mid \mathbf{e}_2 \mid \cdots \mid \mathbf{e}_{p_k}]$. Then $A_k = S_k(D_k + \varepsilon E_k)S_k^{-1} \in CS_{\omega_k}$, $k = 1, \ldots, p_0$, and

$$A = \begin{bmatrix} A_1 & & & \\ & A_2 & & \\ & & \ddots & \\ & & & A_{p_0} \end{bmatrix}.$$

From *i*), there exists an appropriate vector $\mathbf{q}_k^T = (q_{k0}, q_{kp_1}, \ldots, q_{kp_k})$, with $\sum_{j=0}^{p_k} q_{kj} = 0$, such that $A_k + \mathbf{e}\mathbf{q}_k^T$ is nonnegative for each $k = 1, \ldots, p_0$. From Lemma 1, $A_k + \mathbf{e}\mathbf{q}_k^T$ has the prescribed elementary divisors. Thus, from Theorem 3, $M = A + XC$ is nonnegative with spectrum Λ and from Lemma 4, M has the prescribed elementary divisors.■

Example 3 *Let* $\Lambda = \{7, 1, -2, -2, -1 + 3i, -1 - 3i\}$. *We construct a nonnegative matrix A with spectrum* Λ *and with elementary divisors*

$$(\lambda-7), (\lambda-1), (\lambda+2)^2, (\lambda+1-3i), (\lambda+1+3i).$$

Consider the partition $\Lambda = \Lambda_0 \cup \Lambda_1 \cup \Lambda_2 \cup \Lambda_3$ *with*

$$\Lambda_0 = \{7, -1+3i, -1-3i\}; \Lambda_1 = \{-2, -2\}; \Lambda_2 = \{1\}; \Lambda_3 = \emptyset \text{ and}$$
$$\Gamma_1 = \{4, -2, -2\}; \Gamma_2 = \{1, 1\}; \Gamma_3 = \{0\}.$$

The matrix

$$B = \begin{bmatrix} 4 & 0 & 3 \\ \frac{34}{7} & 1 & \frac{8}{7} \\ 0 & 7 & 0 \end{bmatrix}$$

has the spectrum Λ_0 *with diagonal entries* 4, 1, 0. *Then, from Theorem 7, we have*

$$A=\begin{bmatrix}0&2&2&0&0&0\\3&0&1&0&0&0\\2&2&0&0&0&0\\0&0&0&1&0&0\\0&0&0&0&1&0\\0&0&0&0&0&0\end{bmatrix}+\begin{bmatrix}1&0&0\\1&0&0\\1&0&0\\0&1&0\\0&1&0\\0&0&1\end{bmatrix}\begin{bmatrix}0&0&0&0&0&3\\\frac{34}{7}&0&0&0&0&\frac{8}{7}\\0&0&0&7&0&0\end{bmatrix}$$

$$=\begin{bmatrix}0&2&2&0&0&3\\3&0&1&0&0&3\\2&2&0&0&0&3\\\frac{34}{7}&0&0&1&0&\frac{8}{7}\\\frac{34}{7}&0&0&0&1&\frac{8}{7}\\0&0&0&7&0&0\end{bmatrix}$$

with the prescribed spectrum and elementary divisors.

4. NIEDP FOR GENERALIZED DOUBLY STOCHASTIC MATRICES

This section presents the necessary conditions for the existence of nonnegative generalized doubly stochastic matrices with specified elementary divisors. Specifically, we demonstrate a method to convert a generalized stochastic matrix, which may not be nonnegative, with specified elementary divisors, into a nonnegative (positive) generalized doubly stochastic matrix. This transformation results in an increase in the Perron eigenvalue, while leaving the other elementary divisors unaltered. This is what the following result does:

Theorem 8 [11] *Let* $\Lambda = \{\lambda_1, \ldots, \lambda_n\}$ *be a list of complex numbers, with* $\bar{\Lambda}=\Lambda$, $\sum_{i=1}^{n}\lambda_i>0$, $\lambda_1 > |\lambda_i|$, $i = 2, \ldots, n$. *where* $\lambda_1, \ldots, \lambda_p$ *are real and* $\lambda_{p+1}, \ldots, \lambda_n$ *are complex nonreal numbers. Let*

$$(\lambda-\lambda_1),(\lambda-\lambda_2)^{n2},\ldots,(\lambda-\lambda_k)^{nk},n_2+\cdots+n_k=n-1,$$

the prescribed elementary divisors (prescribed JCF). Let M be as defined in (4) and let $m_k = -\min\{0, \mathrm{Re}\lambda_k, \mathrm{Im}\lambda_k\}$, $k = 2, \ldots, n$. *If each of the following statements hold:*

$$i)\ \lambda_1 \geq M + \sum_{k=2}^{n} m_k,$$

$$ii)\ \lambda_1 > Re\lambda_k + Im\lambda_k + n(m_k),\ k = 2, \ldots, n,$$

$$iii)\ Re\lambda_k + Im\lambda_k - \frac{t}{n}\varepsilon < \frac{1}{n}\sum_{k=1}^{n}\lambda_k,\ k = 2,\ldots,n,$$

where ε < 0, with

$$|\varepsilon| \leq \min\left\{ \min_{2\leq k\leq p} |\lambda_k|, \min_{p+1\leq k\leq n} |Im\lambda_k|, \frac{1}{t}\sum_{i=1}^{n} \lambda_i \right\},$$

for real $\lambda_k \neq 0$, $k = 2, \ldots, p$, or

$$|\varepsilon| \leq \min\left\{ \min_{p+1\leq k\leq n} |Im\lambda_k|, \frac{1}{t}\sum_{i=1}^{n} \lambda_i \right\}$$

if $\lambda_k = 0$, with λ_k being the zero of an elementary divisor $(\lambda-\lambda_k)^{n_k}$, $n_k \geq 2$, and t being the total number of times that ε appears in the prescribed JCF, in certain positions $(i, i + 1)$, $i = 2, \ldots, n - 1$, then there exists a nonnegative generalized doubly stochastic matrix with the prescribed elementary divisors.

Proof. Let λ_i be real, $i = 2, \ldots, p$, and let $x_j = Re\lambda_j$ and $y_j = Im\lambda_j$, $j = p + 1, \ldots, n - 1$. Consider the initial matrix $A \in CS_{\lambda 1}$,

$$A = \begin{bmatrix}
\lambda_1 & & & & & & & & & \\
\vdots & \ddots & & & & & & & & \\
\lambda_1-\lambda_l-\varepsilon & & \lambda_l & \varepsilon & & & & & & \\
\lambda_1-\lambda_l & & & \lambda_l & \ddots & & & & & \\
\vdots & & & & \ddots & \ddots & & & & \\
\lambda_1-x_h-y_h & & & & & x_h & y_h & & & \\
\lambda_1-x_h+y_h-\varepsilon & & & & & -y_h & x_h & \varepsilon & & \\
\lambda_1-x_h-y_h & & & & & & & x_h & y_h & \\
\lambda_1-x_h+y_h & & & & & & & -y_h & x_h & \\
\vdots & & & & & & & & & \ddots \\
\lambda_1-x_{n-1}+y_{n-1} & & & & & & & & -y_{n-1} & x_{n-1}
\end{bmatrix},$$

which has the desired *JCF*. First, from Theorem 2, we can transform *A* into a nonnegative matrix $A' \in CS_{\lambda 1}$. Let

$$r_k = m_k, \; k = 2, \ldots, n \quad \text{and}$$

$$\mathbf{r}^T = \left(-\sum_{k=2}^{n} r_k, r_2, r_3, \ldots, r_n \right).$$

Then from condition *i*), the matrix $A' = A + \mathbf{e}\mathbf{r}^T$ is nonnegative, $A' \in CS_{\lambda 1}$, and from Lemma 1, it has same elementary divisors as *A*. We now describe how to perturb *A'* to make it nonnegative generalized doubly stochastic. Define

$$\mathbf{q}^T = \left(-\sum_{k=2}^{n} q_k, q_2, q_3, \ldots, q_n \right),$$

with

$$q_k = \frac{1}{n}(\lambda_1 - Re\lambda_k - Im\lambda_k - nr_k - \varepsilon)$$

if the kth column of A has ε above $Im\lambda_k$ or above a real λ_k, and

$$q_k = \frac{1}{n}(\lambda_1 - Re\lambda_k - Im\lambda_k - nr_k), k = 2,\ldots,n,$$

otherwise. Let $B = A' + \mathbf{eq}^T$. Again by Lemma 1, B has the same elementary divisors as A' (which are the same of A). From *ii)* it is clear that all entries of B, on columns 2 to n, are nonnegative. For the (1, 1) entry of B, we have, from (7) or (8), and assuming that ε appears a total of t times in positions $(i, i + 1)$, that

$$\lambda_1 - \sum_{k=2}^{n} r_k - \sum_{k=2}^{n} q_k = \lambda_1 - \sum_{k=2}^{n} r_k - \frac{1}{n}\sum_{k=2}^{n}(\lambda_1 - Re\lambda_k - Im\lambda_k - nr_k) + \frac{1}{n}t\varepsilon$$

$$= \frac{1}{n}\left(\sum_{k=1}^{n}\lambda_k + t\varepsilon\right) \geq 0.$$

In position $(k, 1)$, $k = 2, \ldots, n$, we have from condition *iii)* that

$$\lambda_1 - Re\lambda_k - Im\lambda_k - \sum_{k=2}^{n} r_k - \sum_{k=2}^{n} q_k = \frac{1}{n}\left(\sum_{k=1}^{n}\lambda_k + t\varepsilon\right) - Re\lambda_k - Im\lambda_k \geq 0.$$

Thus, all entries in B are nonnegative. Now we show that B, $B^T \in CS_{\lambda 1}$: It is clear that $B \in CS_{\lambda 1}$: $B\mathbf{e}=(A'+\mathbf{eq}^T)\mathbf{e}=\lambda_1\mathbf{e}$. Then, since the row sums are each λ_1, and from the way in which the q_k were defined, each the columns 2, ..., n have column sum λ_1. The first column sum is also λ_1.

Example 4 *Let* $\Lambda = \{\lambda_1, 2, 2, -1 + i, -1 - i, -1 + i, -1 - i\}$. *We want to construct a nonnegative generalized doubly stochastic matrix with elementary divisors* $(\lambda - \lambda_1)$, $(\lambda + 2)^2$, $((\lambda + 1)^2 + 1)^2$. *We start with the* 7×7 *initial matrix* $A \in CS_{\lambda 1}$, *and the desired JCF:*

$$A = \begin{bmatrix} \lambda_1 & 0 & 0 & 0 & 0 & 0 & 0 \\ \lambda_1 - 2 - \varepsilon & 2 & \varepsilon & 0 & 0 & 0 & 0 \\ \lambda_1 - 2 & 0 & 2 & 0 & 0 & 0 & 0 \\ \lambda_1 & 0 & 0 & -1 & 1 & 0 & 0 \\ \lambda_1 + 2 - \varepsilon & 0 & 0 & -1 & -1 & \varepsilon & 0 \\ \lambda_1 & 0 & 0 & 0 & 0 & -1 & 1 \\ \lambda_1 + 2 & 0 & 0 & 0 & 0 & -1 & -1 \end{bmatrix}.$$

We apply Theorem 2 to transform A into a nonnegative matrix $A'=A+\mathbf{er}^T \in CS_{\lambda 1}$, *where for* $\varepsilon = -1$,

$$\mathbf{r}^T = \left(-\sum_{k=2}^{7} r_k, r_2, \ldots, r_7 \right) = (-5, 0, 1, 1, 1, 1, 1)$$

Then we obtain

$$A' = A + \mathbf{e}\mathbf{r}^T = \begin{bmatrix} \lambda_1 - 5 & 0 & 1 & 1 & 1 & 1 & 1 \\ \lambda_1 - 6 & 2 & 0 & 1 & 1 & 1 & 1 \\ \lambda_1 - 7 & 0 & 3 & 1 & 1 & 1 & 1 \\ \lambda_1 - 5 & 0 & 1 & 0 & 2 & 1 & 1 \\ \lambda_1 - 2 & 0 & 1 & 0 & 0 & 0 & 1 \\ \lambda_1 - 5 & 0 & 1 & 1 & 1 & 0 & 2 \\ \lambda_1 - 3 & 0 & 1 & 1 & 1 & 0 & 0 \end{bmatrix}.$$

which, by Lemma 1 is a nonnegative generalized stochastic matrix for all $\lambda_1 \geq 7$, *with the same elementary divisors as A. We now perturb A' to make it nonnegative generalized doubly stochastic. Define*

$$\mathbf{q}^T = \left(-\sum_{k=2}^{7} q_k, q_2, \ldots, q_7 \right), \text{with } q_1 = -\frac{1}{7}(6\lambda_1 - 33)$$

$$q_2 = \frac{1}{7}(\lambda_1 - 2), q_3 = \frac{1}{7}(\lambda_1 - 8), q_4 = \frac{1}{7}(\lambda_1 - 7)$$

$$q_5 = \frac{1}{7}(\lambda_1 - 5), q_6 = \frac{1}{7}(\lambda_1 - 7), q_7 = \frac{1}{7}(\lambda_1 - 4)$$

Then $B = A' + \mathbf{e}\mathbf{q}^T$ *is nonnegative generalized doubly stochastic, with same elementary divisors as A, provided that*

$$\lambda_1 - 7 - \frac{1}{7}(6\lambda_1 - 33) \geq 0. \text{ that is } \lambda_1 \geq 16.$$

The following result extends Theorem 8. A proof, which is built on the basis of Theorem 3, can be found in [11, Theorem 3.1]. The constructive nature of the proof allows us to compute a solution matrix.

Theorem 9 [11] *Let* $\Lambda = \{\lambda_1, \lambda_2, \ldots, \lambda_n\}$ *be a list of complex numbers with* $\bar{\Lambda} = \Lambda$, $\sum_{i=1}^{n} \lambda_i > 0$, $\lambda_1 > |\lambda_i|$, $i = 2, \ldots, n$. *Suppose there exists a partition* $\Lambda = \Lambda_0 \cup \Lambda_1 \cup \cdots \cup \Lambda_{p_0}$ *with*

$$\Lambda_0 = \{\lambda_{01}, \lambda_{02}, \ldots, \lambda_{0p_0}\}, \ \lambda_{01} = \lambda_1$$

$$\Lambda_k = \{\lambda_{k1}, \lambda_{k2}, \ldots, \lambda_{kp_k}\}, p_k = p, k = 1, \ldots, p_0,$$

where the lists Λ_k, $k = 1, \ldots, p_0$, have cardinality p, in such a way that:

- *For each $k = 1, \ldots, p_0$, there exists a list*

$$\Gamma_k = \{\omega_k, \lambda_{k1}, \ldots, \lambda_{kp_k}\}, 0 < \omega_k < \lambda_1$$

which is realizable by a nonnegative (positive) matrix A_k, with A_k, with A_k, $A_k^T \in CS_{\omega_k}$, and with prescribed elementary divisors

$$(\lambda - \omega_k), (\lambda - \lambda_{k1})^{n_{k1}}, \ldots, (\lambda - \lambda_{kj})^{n_{kj}}, n_{k1} + \cdots + n_{kj} = p_k,$$

- *There exists a $p_0 \times p_0$ nonnegative (positive) matrix $B = (b_{ij})_{i,j=1}^{n}$ such that B, $B^T \in CS_{\lambda 1}$, with spectrum Λ_0 and diagonal entries $\omega_1, \omega_2, \ldots, \omega_{p0}$, and with certain of the prescribed elementary divisors.*

Then there exists a nonnegative (positive) matrix A, such that A, $A^T \in CS_{\lambda 1}$, with spectrum Λ and with the prescribed elementary divisors associated to the lists Λ_k.

Example 5 *Let $\Lambda = \{12, 5, 2, 2, 0, -1, -1, -2+i, -2-i\}$. To construct a positive generalized doubly stochastic matrix A with elementary divisors*

$$(\lambda-12), (\lambda-5), (\lambda-2)^2, \lambda, (\lambda+1)^2, \lambda^2+4\lambda+5$$

we take the partition

$$\Lambda_0 = \{12, 5, 0\}, \Gamma_1 = \{7, 2, 2\}, \Gamma_2 = \{6, -2+i, -2-i\}, \Gamma_3 = \{4, -1, -1\}$$

and we compute the positive generalized doubly stochastic matrices

$$A_1 = \frac{1}{3}\begin{bmatrix} 10 & 5 & 6 \\ 7 & 11 & 3 \\ 4 & 5 & 12 \end{bmatrix}, A_2 = \frac{1}{3}\begin{bmatrix} 2 & 9 & 7 \\ 5 & 3 & 10 \\ 11 & 6 & 1 \end{bmatrix}, A_3 = \frac{1}{3}\begin{bmatrix} 1 & 5 & 6 \\ 7 & 2 & 3 \\ 4 & 5 & 3 \end{bmatrix}$$

with spectra $\Gamma_1, \Gamma_2, \Gamma_3$, and elementary divisors $(\lambda - 7)$, $(\lambda - 2)^2$; $(\lambda - 6)$, $\lambda^2 + 4\lambda + 5$; and $(\lambda - 4)$, $(\lambda + 1)^2$, respectively. Moreover, we compute

$$B = \begin{bmatrix} 7 & 1 & 4 \\ 2 & 6 & 4 \\ 3 & 5 & 4 \end{bmatrix} \text{ with spectrum } \Lambda_0, \text{ and } C = \begin{bmatrix} 0 & 1 & 4 \\ 2 & 0 & 4 \\ 3 & 5 & 0 \end{bmatrix}.$$

Finally

$$XCX^T = \frac{1}{3}\begin{bmatrix} 0 & 0 & 0 & 1 & 1 & 1 & 4 & 4 & 4 \\ 0 & 0 & 0 & 1 & 1 & 1 & 4 & 4 & 4 \\ 0 & 0 & 0 & 1 & 1 & 1 & 4 & 4 & 4 \\ 2 & 2 & 2 & 0 & 0 & 0 & 4 & 4 & 4 \\ 2 & 2 & 2 & 0 & 0 & 0 & 4 & 4 & 4 \\ 2 & 2 & 2 & 0 & 0 & 0 & 4 & 4 & 4 \\ 3 & 3 & 3 & 5 & 5 & 5 & 0 & 0 & 0 \\ 3 & 3 & 3 & 5 & 5 & 5 & 0 & 0 & 0 \\ 3 & 3 & 3 & 5 & 5 & 5 & 0 & 0 & 0 \end{bmatrix}, \text{ and}$$

$$A = \begin{bmatrix} A_1 & & \\ & A_2 & \\ & & A_3 \end{bmatrix} + XCX^T$$

is positive generalized doubly stochastic matrix with the prescribed elementary divisors.

5. NIEDP FOR PERSYMMETRIC MATRICES

In this section, we consider the *NIEDP* for persymmetric matrices. Persymmetric matrices are common in physical sciences and engineering. They arise, for instance, in the control of mechanical and electric vibrations, where the eigenvalues of the Gram matrix, which is symmetric and persymmetric, play an important role [22]. As the superscript T, in A^T, denotes the transpose of A, the superscript F, in A^F, denotes the flip-transpose of A, which flips A across its skew-diagonal. If $A = (a_{ij})_{mn}$, then $A^F = (a_{n-j+1,m-i+1})_{nm}$. A matrix A is said to be *persymmetric* if $A^F = A$, that is, if it is symmetric across its lower-left to upper-right diagonal. Let J be the $n \times n$ matrix with ones along its skew-diagonal and zeroes elsewhere, that is, $J = [\mathbf{e}_n|\mathbf{e}_{n-1}| \cdots |\mathbf{e}_1]$. Then

$$J^T = J^F = J, J^2 = I, A^F = JA^TJ,$$

and the following properties are straightforward:

$$\left(A^F\right)^F = A,\ \left(A^T\right)^F = \left(A^F\right)^T,\ \left(A+B\right)^F = A^F + B^F,\ \left(AB\right)^F = B^F A^F$$

Proposition 10 *If A and B are persymmetric matrices, then i)* $aA + bB$, *with* $a, b \in \mathbb{R}$,

◉ A^{-1}, *if* A^{-1} *exists, and iii)* A^T, *are also persymmetric.*

Proof. *i)* $(aA + bB)^F = aA^F + bB^F = aA + bB$.

◉ $I = A^{-1}A$ implies $A^F(A^{-1})^F = A(A^{-1})^F = I$. Hence, $(A^{-1})^F = A^{-1}$.

◉ $(A^T)^F=(A^F)^T=A^T$.

In this section, we give sufficient conditions for the existence of nonnegative persymmetric matrices with prescribed elementary divisors. Our result generates an algorithmic procedure to compute a solution matrix. We also show that companion matrices are similar to persymmetric ones. As a consequence, we show that any realizable list of complex numbers $\Lambda = \{\lambda_1, ..., \lambda_n\}$, with $\mathrm{Re}\lambda_i \le 0$, $i = 2, ..., n$, is in particular realizable by a nonnegative persymmetric matrix.

In [23], the authors develop the following persymmetric version of Rado's Theorem 3, which allow us to obtain sufficient conditions for the existence of a persymmetric nonnegative matrix with prescribed spectrum.

Theorem 11 [23] *Let A be an $n \times n$ persymmetric matrix with spectrum*

$\Lambda = \{\lambda_1, \lambda_2, ..., \lambda_n\}$, and for some $r \le n$, let $\{\mathbf{x}_1, \mathbf{x}_2, ..., \mathbf{x}_r\}$ be a set of eigenvectors of A corresponding to $\lambda_1, ..., \lambda_r$, respectively. Let X be the $n \times r$ matrix with i – th column $\mathbf{x}_i$ and rank(X) = r. Let $\Omega = diag\{\lambda_1, ..., \lambda_r\}$, and let C be an $r \times r$ persymmetric matrix. Then, the matrix $A + XCX^F$ is persymmetric with eigenvalues $\mu_1, \mu_2, ..., \mu_r, \lambda_{r+1}, ..., \lambda_n$, where $\mu_1, \mu_2, ..., \mu_r$ are eigenvalues of $B = \Omega + CX^F X$.

Let $\Lambda = \{\lambda_1, \lambda_2, ..., \lambda_n\}$ be a list of complex numbers, which can be partitioned as

$$\left.\begin{aligned} &\Lambda = \Lambda_0 \cup \Lambda_1 \cup \cdots \cup \Lambda_{\frac{p_0}{2}} \cup \Lambda_{\frac{p_0}{2}} \cup \cdots \cup \Lambda_1, \text{for even } p_0, \\ &\text{with} \qquad \Lambda_0 = \{\lambda_{01}, \lambda_{02}, \ldots, \lambda_{0p_0}\}, \ \lambda_{01} = \lambda_1, \\ &\Lambda_k = \{\lambda_{k1}, \lambda_{k2}, \ldots, \lambda_{kp_k}\}, \ k = 1, 2, \ldots, \frac{p_0}{2}, \end{aligned}\right\}$$

where some of the lists Λ_k can be empty. For each list Λ_k, we associate the list

$$\Gamma_k = \{\omega_k, \lambda_{k1}, \lambda_{k2}, \ldots, \lambda_{kp_k}\}, 0 \le \omega_k \le \lambda_1,$$

which is realizable by a $(p_k + 1) \times (p_k + 1)$ nonnegative matrix A_k. In particular, A_k can be chosen as $A_k \in CS_{\omega k}$.

Let the $n \times p_0$ matrix

$$X = \begin{bmatrix} \mathbf{x}_1 & & & & \\ & \ddots & & & \\ & & \mathbf{x}_{\frac{p_0}{2}} & & \\ & & & \mathbf{y}_{\frac{p_0}{2}} & & \\ & & & & \ddots & \\ & & & & & \mathbf{y}_1 \end{bmatrix}, \text{ for even } p_0.$$

where, $\mathbf{x}_k = \mathbf{e} = [1, 1, ..., 1]^T$ and $\mathbf{y}_k = [y_{k1}, \ldots, y_{k(p_k+1)}]^T \ge 0$ are, respectively, the Perron eigenvectors of $A_k^F, k = 1, 2, \ldots, \frac{p_0}{2}$.. It was shown in [23] that in the case of even p_0, $X^F X$ is diagonal persymmetric nonnegative matrix with positive maindiagonal.ToapplyTheorem11,weneedtofinda$p_0 \times p_0$persymmetricnonnegative matrix C, where $B = \Omega + CX^F X$ is persymmetric nonnegative with eigenvalues μ_1, $\mu_2,\ldots,\mu_{p0}$ (the new eigenvalues) and diagonal entries $\omega_1, \ldots, \omega_{\frac{p_0}{2}}, \omega_{\frac{p_0}{2}}, \ldots, \omega_1$ (the former eigenvalues), and $\Omega=\text{diag}\left\{\omega_1, \ldots, \omega_{\frac{p_0}{2}}, \omega_{\frac{p_0}{2}}, \ldots, \omega_1\right\}$. Thus, since $X^F X$ is nonsingular, we may compute $C = (B - \Omega)(X^F X)^{-1}$. However, although both matrices, $(B - \Omega)$ and $(X^F X)^{-1}$ are persymmetric, their product need not to be persymmetric. It was shown in [23, Lemma 3.1] that we may transform the matrix X into a matrix $\widetilde{X}$ in such a way that $\widetilde{X}^F \widetilde{X} = sI$, $s > 0$, , with the columns of $\widetilde{X}$ still being eigenvectors of A. Now we state the main result of this section, which was proven in [10]. We only consider the case of even p_0. The odd case is similar.

Theorem 12 [10]*Let $\Lambda = \{\lambda_1, \lambda_2, \ldots, \lambda_n\}$ be a list of complex numbers with $\Lambda = \overline{\Lambda}$, $\lambda_1 \geq |\lambda_i|$, $i = 2, 3, \ldots, n$, and $\sum_{i=1}^{n} \lambda_i \geq 0$. Suppose there exists a partition of Λ as in (9), such that the following conditions are satisfied:*

- *For each k=1,2,..., $\frac{p_0}{2}$, , there exists a nonnegative matrix with spectrum*

$$\Gamma_k = \left\{\omega_k, \lambda_{k1}, \lambda_{k2}, \ldots, \lambda_{kp_k}\right\}, 0 \leq \omega_k \leq \lambda_1,$$

and with certain prescribed elementary divisors.

- *There exists a persymmetric nonnegative matrix of order p_0, with spectrum Λ_0, diagonal entries $\omega_1, \omega_2, \ldots, \omega_{\frac{p_0}{2}}, \omega_{\frac{p_0}{2}}, \ldots, \omega_2, \omega_1$, and certain prescribed elementary divisors.*

Then Λ is realizable by an $n \times n$ persymmetric nonnegative matrix with the prescribed elementary divisors.

Proof. From *i*), let A_k be a nonnegative matrix with spectrum Γ_k, and certain prescribed elementary divisors, $k = 1, 2, \ldots, \frac{p_0}{2}$. We may assume that $A_k \in CS_{\omega k}$. Then $A_k \mathbf{e} = \omega_k \mathbf{e}$, with $\mathbf{e} = (1, 1, \ldots, 1)^T$. The matrix

$$A = A_1 \oplus A_2 \oplus \cdots \oplus A_{\frac{p_0}{2}} \oplus A^F_{\frac{p_0}{2}} \oplus \cdots \oplus A_2^F \oplus A_1^F$$

is persymmetric nonnegative with spectrum $\Gamma = \Gamma_1 \cup \cdots \cup \Gamma_{\frac{p_0}{2}} \cup \Gamma_{\frac{p_0}{2}} \cup \cdots \cup \Gamma_1$, and the prescribed elementary divisors associated to the lists Λ_k. Let X (or $\widetilde{X}$) be the matrix in (11) such that $X^F X = sI_{p_0}$, $s > 0$ (or $\widetilde{X}^F \widetilde{X} = sI_{p_0}$, $s > 0$). Now, from *ii*) let

$B = \Omega + CX^F X = \Omega + sCI_{p_0}$ be the $p_0 \times p_0$ persymmetric nonnegative matrix with spectrum Λ_0 and diagonal entries $\omega_1, \ldots, \omega_{\frac{p_0}{2}}, \omega_{\frac{p_0}{2}}, \ldots, \omega_1$. Then $C = \frac{1}{s}(B - \Omega)$ is persymmetric nonnegative, where $\Omega = diag\left\{\omega_1, \ldots, \omega_{\frac{p_0}{2}}, \omega_{\frac{p_0}{2}}, \ldots, \omega_1\right\}$. Thus, from Theorem 11, $M = A + XCX^F$ is persymmetric with spectrum Λ, and with the prescribed elementary divisors. Since A, X, and C are nonnegative, then M is also nonnegative.

In [10], the authors show the following result:

Theorem 13 [10] *Let*

$$C = \begin{bmatrix} 0 & 1 & 0 & \cdots & 0 \\ 0 & 0 & 1 & \ddots & 0 \\ \ddots & \ddots & \ddots & \ddots & \vdots \\ 0 & \ddots & \ddots & \ddots & 1 \\ c_{n-1} & \cdots & c_2 & c_1 & c_0 \end{bmatrix}$$

be the companion matrix of the polynomial $p(x) = x^n + \sum_{k=1}^{n} c_{k-1} x^{n-k}$. *Then* **C** *is similar to a persymmetric matrix* **P**. *In particular, if* **C** *is nonnegative,* **P** *is also nonnegative.*

In [21], Laffey and Šmigoc solve the NIEP in the left half plane. That is, they give necessary and sufficient conditions for the existence of a nonnegative matrix with prescribed complex spectrum $\Lambda = \{\lambda_1, \lambda_2, \ldots, \lambda_n\}$ satisfying $\text{Re}\lambda_i \leq 0$, $i = 2, \ldots, n$. Here we show that a realizable list of complex numbers, with $\text{Re}\lambda_i \leq 0$, $i = 2, \ldots, n$, is in particular realizable by a nonnegative persymmetric matrix.

Theorem 14 [10] *Let* $\Lambda = \{\lambda_1, \lambda_2, \ldots, \lambda_n\}$, *with* $\text{Re}\lambda_i \leq 0$, $i = 2, \ldots, n$, *be a realizable list of complex numbers. Then* Λ *is also realizable by a nonnegative persymmetric matrix.*

Proof. If Λ is realizable, then from the result in [21], Λ is the spectrum of a matrix of the form $\mathbf{C} + \alpha I$, where $\mathbf{C}$ is a nonnegative companion matrix with $tr(\mathbf{C})=0$. Then from Theorem 13, C is similar to a nonnegative persymmetric matrix P with eigenvalues $\lambda_1 - \alpha, \lambda_2 - \alpha, \ldots, \lambda_n - \alpha$. Hence $P+\alpha I$ is nonnegative persymmetric with spectrum Λ.

To apply Theorem 12, we need to know conditions under which there exists a $p_0 \times p_0$ persymmetric nonnegative matrix with spectrum Λ_0 and diagonal entries $\omega_1, \omega_2, \ldots, \omega_{\frac{p_0}{2}}, \omega_{\frac{p_0}{2}}, \ldots, \omega_2, \omega_1$. Following a result due to Farahat and Ledermann [24, Theorem 2.1], we have **Theorem 15** *[24] Let* $\Lambda = \{\lambda_1, \lambda_2, \ldots, \lambda_n\}$ *be a list of complex numbers and let* $\{a_1, a_2, \ldots, a_n\}$ *be a list of nonnegative real numbers such that* $\sum_{i=1}^{n} a_i = \sum_{i=1}^{n} \lambda_i$. *Let* $p(x) = (x - \lambda_1)(x - \lambda_2) \cdots (x - \lambda_n)$ *and*

$$\mu_0 = 1, \mu_1 = (x - a_1), \cdots, \mu_n = (x - a_1)(x - a_2)\cdots(x - a_n),$$

with

$$p(x) = \mu_n + k_1\mu_{n-1} + \cdots + k_{n-1}\mu_1 + k_n.$$

If $k_1, k_2, ..., k_n$ are all nonpositive, then there exists an $n \times n$ nonnegative matrix, with spectrum Λ and with diagonal entries $a_1, a_2, ..., a_n$.

Proof. The polynomials $\mu_0, \mu_1, ..., \mu_n$ constitute a basis in the space of polynomials of degree less than or equal to n. Equating the coefficients of x^{n-1} in both sides in (12), we obtain $k_1 = 0$. Let the matrix

$$A = \begin{bmatrix} a_1 & 1 & 0 & \cdots & 0 \\ 0 & a_2 & 1 & \ddots & \vdots \\ \ddots & \ddots & \ddots & \ddots & 0 \\ \ddots & \ddots & 0 & \ddots & 1 \\ -k_n & \cdots & -k_3 & -k_2 & a_n \end{bmatrix}$$

Expanding the determinant of $xI - A$ with respect to the last row it, follows that A has characteristic polynomial $p(x)$ and therefore A has spectrum Λ. Besides, if k_1, ..., k_n are all nonpositive, A is nonnegative.

Example 6 *Let $\Lambda = \{9, -3, -3, -3\}$. First we compute a persymmetric nonnegative matrix* P *with spectrum Λ and with elementary divisors $(\lambda - 9), (\lambda + 3)^3$. Then $p(x) = (\lambda - 9)(\lambda + 3)^3 = \lambda^4 - 54\lambda^2 - 216\lambda - 243$, and the corresponding matrices* C *(the companion of $p(x)$) and* P *(obtained from Theorem 13) are, respectively,*

$$C = \begin{bmatrix} 0 & 1 & 0 & 0 \\ 0 & 0 & 1 & 0 \\ 0 & 0 & 0 & 1 \\ 243 & 216 & 54 & 0 \end{bmatrix} \rightarrow P = \begin{bmatrix} 0 & 1 & 0 & 0 \\ 27 & 0 & 1 & 0 \\ 108 & 0 & 0 & 1 \\ 972 & 108 & 27 & 0 \end{bmatrix}.$$

Next, we compute a persymmetric matrix P *with spectrum Λ and linear elementary divisors: Then we take the partition*

$$\Lambda_0 = \{9, -3\}, \Lambda_1 = \Lambda_2 = \{-3\} \text{ with } \Gamma_1 = \Gamma_2 = \{3, -3\}.$$

We apply Theorem 12 with

$$A = \begin{bmatrix} 0 & 3 & 0 & 0 \\ 3 & 0 & 0 & 0 \\ 0 & 0 & 0 & 3 \\ 0 & 0 & 3 & 0 \end{bmatrix}, \; X = \begin{bmatrix} \frac{1}{\sqrt{2}} & 0 \\ \frac{1}{\sqrt{2}} & 0 \\ 0 & \frac{1}{\sqrt{2}} \\ 0 & \frac{1}{\sqrt{2}} \end{bmatrix}, \; B = \begin{bmatrix} 3 & 1 \\ 36 & 3 \end{bmatrix},$$

to obtain the persymmetric nonnegative matrix $A + XCX^F$ *with the required properties. Finally, we compute a persymmetric nonnegative matrix* P *with spectrum* Λ *and with elementary divisors* $(\lambda - 9)$, $(\lambda + 3)^2$, $(\lambda + 3)$. *In this case, we take the partition*

$$\Lambda_0 = \{9,-3,-3\}, \Lambda_1 = \Lambda_3 = \varnothing, \Lambda_2 = \{-3\}$$

$$\Gamma_1 = \Gamma_3 = \{0\}, \Gamma_2 = \{3,-3\}$$

$$A_1 = A_3 = [0],\ A_2 = \begin{bmatrix} 0 & 3 \\ 3 & 0 \end{bmatrix}.$$

We compute the matrix B, with spectrum Λ_0 *and diagonal entries* 0, 3, 0 *(in that order), from Theorem* 15, *case* $n = 3$:

$$B = \begin{bmatrix} 0 & 1 & 0 \\ \frac{45}{2} & 3 & 1 \\ 81 & \frac{45}{2} & 0 \end{bmatrix} \text{ with } J(B) = \begin{bmatrix} 9 & 0 & 0 \\ 0 & -3 & 1 \\ 0 & 0 & -3 \end{bmatrix}$$

Then,

$$A = \begin{bmatrix} 0 & 0 & 0 & 0 \\ 0 & 0 & 3 & 0 \\ 0 & 3 & 0 & 0 \\ 0 & 0 & 0 & 0 \end{bmatrix} + \begin{bmatrix} 1 & 0 & 0 \\ 0 & \frac{1}{\sqrt{2}} & 0 \\ 0 & \frac{1}{\sqrt{2}} & 0 \\ 0 & 0 & 1 \end{bmatrix} \begin{bmatrix} 0 & 1 & 0 \\ \frac{45}{2} & 0 & 1 \\ 81 & \frac{45}{2} & 0 \end{bmatrix} \begin{bmatrix} 1 & 0 & 0 & 0 \\ 0 & \frac{1}{\sqrt{2}} & \frac{1}{\sqrt{2}} & 0 \\ 0 & 0 & 0 & 1 \end{bmatrix}$$

$$= \begin{bmatrix} 0 & \frac{1}{2}\sqrt{2} & \frac{1}{2}\sqrt{2} & 0 \\ \frac{45}{4}\sqrt{2} & 0 & 3 & \frac{1}{2}\sqrt{2} \\ \frac{45}{4}\sqrt{2} & 3 & 0 & \frac{1}{2}\sqrt{2} \\ 81 & \frac{45}{4}\sqrt{2} & \frac{45}{4}\sqrt{2} & 0 \end{bmatrix},\ J(A) = \begin{bmatrix} 9 & 0 & 0 & 0 \\ 0 & -3 & 1 & 0 \\ 0 & 0 & -3 & 0 \\ 0 & 0 & 0 & -3 \end{bmatrix}.$$

6. SOME OPEN QUESTIONS

In [3, Theorem 1], Minc proved the following result:

Theorem 16 [3]*Let* $\Lambda = \{\lambda_1, \lambda_2, \ldots, \lambda_n\}$ *be a list of complex numbers, which is realizable by a diagonalizable positive matrix A. Then, for each JCF* J_Λ *associated with* Λ, *there exists a positive matrix B with the same spectrum as A, and with JCF* $J(B) = J_\Lambda$.

According to Minc, the positivity condition is essential in his proof, and it is not known if the result holds without this condition (see [2]). Specifically, it is not known *i)* whether for every positive matrix, there exists a diagonalizable positive matrix with the same spectrum, *ii)* whether for every nonnegative diagonalizable matrix with spectrum $\Lambda = \{\lambda_1, ..., \lambda_n\}$, there exists a nonnegative matrix for each *JCF* associated with Λ. There are many examples which show that the Minc result holds for diagonalizable nonnegative matrices. For instance, all diagonalizable nonnegative matrices with spectrum Λ given in Section 2, Theorems 4 and 5, and Corollaries 1 and 2, give rise to nonnegative matrices for each one of the possible *JCF* associated with Λ. However, we do not know if the Minc result holds for a general diagonalizable nonnegative matrix.

The task of identifying a nonnegative matrix with a predetermined spectrum and diagonal elements is a challenging unresolved issue. This problem is not only significant on its own but also essential for the application of various versions of Rado's Theorem to both the NIEP and the NIEDP difficulties. Necessary and sufficient conditions are only known for $n \leq 3$.

REFERENCES

1. MT Chu, GH Golub (2005) Inverse eigenvalue problems: theory, algorithms and applications, Oxford University Press, New York.
2. H. Minc (1988) Nonnegative Matrices, John Wiley & Sons, New York.
3. H. Minc (1981) Inverse elementary divisors problem for nonnegative matrices. In: Proc. AMS 83, No 4, 665–669.
4. H. Minc (1982) Inverse elementary divisors problem for doubly stochastic matrices. In: Linear and Multilinear Algebra 11 121–131.
5. C. Marijuán, M. Pisonero, RL Soto (2007) A map of sufficient conditions for the real nonnegative inverse eigenvalue problem. In: Linear Algebra Appl. 426 (2007) 690–705.
6. RL Soto, J. Ccapa (2008) Nonnegative matrices with prescribed elementary divisors. In: Electronic Journal of Linear Algebra 17 287–303.
7. J. Ccapa, RL Soto (2009) On spectra perturbation and elementary divisors of positive matrices. In: Electronic Journal of Linear Algebra 18 462–481.
8. J. Ccapa, RL Soto (2010) On elementary divisors perturbations of nonnegative matrices. In: Linear Algebra Appl. 432 546–555.
9. RL Soto, RC Daz, H. Nina, M. Salas (2013) Nonnegative matrices with prescribed spectrum and elementary divisors. In: Linear Algebra Appl. 439 3591–3604.
10. RL Soto, AI Julio, M. Salas (2015) Nonnegative persymmetric matrices with prescribed elementary divisors. In: Linear Algebra Appl. 483 139–157.
11. RL Soto, E. Valero, M. Salas, H. Nina (2015) Nonnegative generalized doubly stochastic matrices with prescribed elementary divisors. In: Electronic Journal of Linear Algebra 30 704–720.

12. RC Daz, RL Soto (2016) Nonnegative inverse elementary divisors problem in the left half plane. In: Linear and Multilinear Algebra 64 (N° 2) 258–268.

13. A. Brauer (1952) Limits for the characteristic roots of a matrix. IV: applications to stochastic matrices. In: Duke Math. J., 19 75–91.

14. H. Perfect (1955) Methods of constructing certain stochastic matrices II. In: Duke Math. J. 22 305–311.

15. RL Soto, O. Rojo (2006) Applications of a Brauer Theorem in the nonnegative inverse eigenvalue problem. In: Linear Algebra Appl. 416 (2–3) (2006) 844–856.

16. R. Reams (1996) An inequality for nonnegative matrices and the inverse eigenvalue problem. In: Linear and Multilinear Algebra 41 367–375.

17. H. Perfect (1953) Methods of constructing certain stochastic matrices. In: Duke Math. J. 20 395–404.

18. HR Suleimanova (1949) Stochastic matrices with real characteristic values. In: Dokl. Akad. Nauk SSSR 66 343–345.

19. A. Borobia, J. Moro, RL Soto (2004) Negativity compensation in the nonnegative inverse eigenvalue problem. In: Linear Algebra Appl. 393 73–89.

20. H. Šmigoc (2004) The inverse eigenvalue problem for nonnegative matrices. In Linear Algebra Appl. 393 365–374.

21. TJ Laffey, H. Šmigoc (2006) Nonnegative realization of spectra having negative real parts. In: Linear Algebra Appl. 416 148–159.

22. RM Reid (1997) Some eigenvalues properties of persymmetric matrices. In: SIAM Rev. 39 313–316.

23. AI Julio, RL Soto (2015) Persymmetric and bisymmetric nonnegative inverse eigenvalue problem. In: Linear Algebra Appl. 469 130–152.

24. HK Farahat, W. Ledermann (1958) Matrices with prescribed polynomials. In: Proc. Edinburg Mat. Soc 11 143–146.

05 ADVANCES IN NONLINEAR INVERSE SCATTERING IN 2D: THEORY AND NUMERICS

1. INTRODUCTION: AN OVERVIEW

We deal with the generalized nonlinear Schrödinger equation:

$$i\frac{\partial}{\partial t}E(x,t) = -\Delta E(x,t) + h(x,|E|)E(x,t),$$

where E denotes the electromagnetic field in two-dimensional case, Δ is the two-dimensional Laplacian and h describes in a general form the nonlinear contribution to the index of refraction. Considering harmonic time dependence $E(x, t) = e^{-i\omega t}u(x)$ with frequency $\omega > 0$, we obtain the steady-state nonlinear equation with fixed energy:

$$-\Delta u(x) + h(x,|u|)u(x) = k^2 u(x), \qquad (1)$$

where $k^2 = \omega$ and fixed, and u denotes the complex amplitude of the field. Concerning the nonlinearity $h(x, s)$, we pose the basic assumptions.

1. $|h(x, s)| \leq c_\rho \alpha(x),\ \alpha \in L^p_\sigma(\mathbb{R}^2),\ 0 \leq s \leq \rho,$
2. $|h(x, s_1) - h(x, s_2)| \leq c'_\rho \beta(x)|s_1 - s_2|,\ \beta \in L^p_\sigma(\mathbb{R}^2),\ 0 \leq s_1, s_2 \leq \rho,$

where c_ρ and c'_ρ are constants and

$$\sigma > 2 - 2/p, 1 < p \leq \infty. \qquad (2)$$

Here $L^p_\sigma(\mathbb{R}^2)$ denotes the weighted Lebesgue space with the norm.

$$||f||_{p,\sigma} = \left(\int_{\mathbb{R}^2} (1+|x|)^{p\sigma}|f(x)|^p dx\right)^{1/p}, \qquad ||f||_p = ||f||_{p,0}.$$

The main practical example (it can be considered as the motivation of this research) of such type equations (1) is the equation of the form:

$$-\Delta u(x) + q_1(x)u(x) + q_2(x)\frac{|u(x)|^2}{1+r|u(x)|^2}u(x) = k^2 u(x), \qquad (3)$$

with real number k, complex valued function $q_1(x) \in L^2$ and real-valued function $q_2(x) \in L^2$, and parameter $r \geq 0$. A particular nonlinearity in (3) of cubic type ($r = 0$) can be met in the context of a Kerr-like nonlinear dielectric film, while the case when $r > 0$ corresponds to the saturation model (see [1–4]).

We consider the inverse scattering problems for (1). For these purposes, we are interested of the scattering solutions to (1), i.e. solutions of the form

$$u(x,k,\theta) = u_0(x,k,\theta) + u_{sc}(x,k,\theta), \qquad (4)$$

where $\theta \in \mathbb{S}^1$, the unit sphere in $\mathbb{R}^2$, $u_0(x, k, \theta) = e^{ik(x,\theta)}$ is the incident wave and $u_{sc}(x, k, \theta)$ is the scattered wave. The scattered wave must satisfy the Sommerfeld radiation condition at infinity:

$$\lim_{r\to\infty} \sqrt{r}\left(\frac{\partial u_{sc}(x,k,\theta)}{\partial r} - iku_{sc}(x,k,\theta)\right) = 0, \qquad r = |x|,$$

for fixed $k > 0$ and uniformly in $\theta \in \mathbb{S}^1$. In that case, these solutions are the unique solutions of the Lippmann-Schwinger equation.

$$u(x,k,\theta) = u_0(x,k,\theta) - \int_{\mathbb{R}^2} G_k^+(|x-y|)h(y,|u(y,k,\theta)|)u(y,k,\theta)dy, \qquad (5)$$

where G_k^+ is the outgoing fundamental solution of the operator $-\Delta - k^2$, i.e. the kernel of the integral operator $(-\Delta - k^2 - i0)^{-1}$. It is equal to

$$G_k^+(|x|) = \frac{i}{4}H_0^{(1)}(|kx|),$$

where $H_0^{(1)}$ denotes the Hankel function of order zero and first kind.

The following main results concerning the direct scattering problem were proved in [5].

Under the basic assumptions and (2) for h, it is proved that for any $\rho > 1$ there is $k_0 > 0$ such that for any $k \geq k_0$ in the ball $B_\rho = \{u \in L^\infty(\mathbb{R}^2) : \|u\|_\infty \leq \rho\}$, there is a unique scattering solution (or the solutions of the form in (4) to (5) which satisfies the condition:

$$||u_{sc}||_\infty \to 0, k \to \infty \qquad (6)$$

uniformly in $\theta \in \mathbb{S}^1$. What is more, the solution is obtained as the limit

$$u(x,k,\theta) = \lim_{j\to\infty} u_j(x,k,\theta),$$

where

$$u_{j+1}(x,k,\theta) = u_0(x,k,\theta) - \int_{\mathbb{R}^2} G_k^+(|x-y|)h(y,|u_j(y,k,\theta)|)u_j(y,k,\theta)dy,$$

for $j = 0, 1, \ldots$ with u_0 as above. Let the function h have the same properties as above, but now with

$$\sigma > \begin{cases} 2 - 2/p, 4/3 \leq p \leq \infty \\ 1/2, 1 < p < 4/3. \end{cases} \qquad (7)$$

Then for fixed $k \geq k_0$, the solution $u(x, k, \theta)$ admits the representation

$$u(x,k,\theta) = e^{ik(x,\theta)} - \frac{1+i}{4\sqrt{\pi k|x|}} e^{ik|x|}A(k,\theta',\theta) + o(|x|^{-1/2}), \qquad |x| \to \infty.$$

The function $A(k, \theta', \theta)$ is called the scattering amplitude and it is defined as

$$A(k,\theta',\theta) = \int_{\mathbb{R}^2} e^{-ik(\theta',y)} h(y, |u(y,k,\theta)|) u(y,k,\theta) dy,$$

where $\theta' = \frac{x}{|x|} \in \mathbb{S}^1$ is the direction of observation. This function A gives us the scattering data for inverse problem. More precisely, the inverse problem that is considered here is to extract some information about the function h by the knowledge of the scattering amplitude A for different sets of scattering data. There are four well-known inverse scattering data sets: (i) the full (scattering) data:

$$D = \{A(k,\theta',\theta): k > 0, \theta', \theta \in \mathbb{S}^1\},$$

(ii) the backscattering data:

$$D_B = \{A(k,\theta',\theta): k > 0, \theta' = -\theta \in \mathbb{S}^1\},$$

(iii) the fixed angle data:

$$D_A = \{A(k,\theta',\theta): k > 0, \theta = \theta_0 \; fixed \; \theta' \in \mathbb{S}^1\}$$

and (iv) the fixed energy data:

$$D_E = \{A(k,\theta',\theta): k = k_0 > 0 \; fixed, \theta', \theta \in \mathbb{S}^1\}.$$

We use the following notations for the direct and inverse Fourier transforms:

$$F(f)(\xi) = \int_{\mathbb{R}^2} e^{i(\xi,y)} f(y) dy \tag{8}$$

$$F^{-1}(f)(x) = \frac{1}{(2\pi)^2} \int_{\mathbb{R}^2} e^{-i(\xi,x)} f(\xi) d\xi, \tag{9}$$

where (ξ, x) denotes the inner product in R2ℝ2, i.e. $(\xi, x) = \xi_1 x_1 + \xi_2 x_2$. By $C > 0$, we denote a generic constant that may change from one step to another. By $H^t(\mathbb{R}^2) = W_2^t(\mathbb{R}^2)$, $t \in \mathbb{R}$ we denote the standard L^2 based Sobolev space. A weighted Sobolev space $W^1_{p,\sigma}(\mathbb{R}^2)$ is defined here by

$$W^1_{p,\sigma}(\mathbb{R}^2) = \{f \in L^p_\sigma(\mathbb{R}^2): \nabla f \in L^p_\sigma(\mathbb{R}^2)\}.$$

The following notation for the characteristic function is used:

$$\chi_A(x) = \begin{cases} 1, x \in A \\ 0, x \notin A. \end{cases}$$

2. INVERSE SCATTERING PROBLEMS

The aforementioned direct scattering hypothesis can also be inverted. The inverse scattering theory treats the function h as unknown and attempts to recover it from the knowledge of the scattering amplitude A for different data. Usually, the model in (1) is probed with one or more incident plane waves u_0 and the resulting scattered waves are measured at a distance. This gives rise to several different scattering data sets which can be used to recover the unknowns.

The inverse backscattering problem for (1) was treated in [6]. Also for (1), the recovery of unknown function h is possible from the full scattering data. In addition to the two-dimensional studies mentioned above, certain particular nonlinear cases of (1) have been investigated in other dimensions too. In one-space dimension, we refer to [7] and the references therein. In higher dimensions $n \geq 3$ we are only aware of [8,9]. Similar problems with formally more general equation but with bounded h are considered in [10] and [11].

Our point of view is that the nonlinearity may contain local singularities in the space coordinate x, and therefore we work in the frame of weighted Lebesgue spaces. These local singularities can be recovered from the scattering amplitude using the method of Born approximation. As a unifying result, we obtain mathematically more general results that have far wider applicability in physical experiments.

Let us set

$$h_0(x) = h(x, 1).$$

In the subsections that follow we consider the inverse problems of recovering information about h_0 from the knowledge of full data D, backscattering data D_B, fixed angle data D_A and fixed energy data D_E.

2.1. Full scattering data

The inverse problem with full data D was investigated in [5]. Here we summarize the main results without proofs.

Theorem 1 (Saito's formula). Under the basic assumptions and (7) for the function h we have,

$$\lim_{k\to\infty} k \int_{\mathbb{S}^1\times\mathbb{S}^1} e^{-ik(\theta-\theta',x)} A(k,\theta',\theta)d\theta d\theta' = 4\pi \int_{\mathbb{R}^2} \frac{h_0(y)}{|x-y|} dy, \qquad x \in \mathbb{R}^2,$$

where the limit is valid in the sense of distributions for $4/3 < p \leq 2$ and pointwise (even uniformly) for $2 < p \leq \infty$.

Corollary 1 (Uniqueness). Let σ be as in (7). Consider the scattering problems for two sets of potentials h and h˜$\tilde{h}$. If the scattering amplitudes coincide for some sequence $k_j \to \infty$ and for all $\theta', \theta \in \mathbb{S}^1$, then

$$h_0(x) = \widetilde{h_0}(x)$$

holds in the sense of distributions for $4/3 < p \leq \infty$.

Corollary 2 (Representation formula). Let σ be as in (7). Then the representation

$$h_0(x) = \lim_{k\to\infty} \frac{k^2}{8\pi^2} \int_{\mathbb{S}^1\times\mathbb{S}^1} e^{-ik(\theta-\theta',x)} A(k,\theta',\theta)|\theta-\theta'|d\theta d\theta'$$

holds in the sense of distributions for $4/3 < p \leq \infty$.

Remark 1. In addition to providing the above results, Saito's formula might be applied numerically too. It can be written as a convolution equation:

$$4\pi h_0 * |x|^{-1} = f,$$

where the function f can be computed from the full scattering data in principle. A numerical inversion of this equation would yield a full recovery of h_0 but this is an open problem as far as we know. What is more, unlike the representation formula above it holds pointwise in the important case of bounded ($p = \infty$) nonlinearities.

We assume that the function $x \mapsto h(x, s)$ is real-valued and recall that

$$A(k,\theta',\theta) = \int_{\mathbb{R}^2} e^{-ik(\theta',y)} h(y,|u|)u(y,k,\theta)dy, \qquad k \geq k_0.$$

For reasons of purely technical nature we define the scattering solutions $u(x, k, \theta)$ for negative k as

$$u(x,k,\theta) = \overline{u(x,-k,\theta)}, \qquad k < 0.$$

These are the unique solutions of the integral equation:

$$u(x,k,\theta) = e^{ik(x,\theta)} - \int_{\mathbb{R}^2} \overline{G_k^+(|x-y|)} h(y,|u|)u(y,k,\theta)dy,$$

provided that h is real-valued. This allows us to extend A to negative $k \leq -k_0$ by

$$A(k,\theta',\theta) = \overline{A(-k,\theta',\theta)}.$$

We also put $A(k, \theta', \theta) = 0$ for $|k| \leq k_0$. Splitting

$$h(y,|u|)u = h_0(x)u_0 + h(y,|u|)u - h_0(x)u_0$$

we have that

$$A(k,\theta',\theta) = \int_{\mathbb{R}^2} e^{ik(\theta-\theta',y)} h_0(y)dy + \int_{\mathbb{R}^2} e^{-ik(\theta',y)} (h(y,|u|)u(y,k,\theta) - h_0(y)u_0)dy$$

or

$$A(k,\theta',\theta) = F(h_0)\big(k(\theta-\theta')\big) + A_R(k,\theta',\theta),$$

where F denotes the Fourier transform (8). Using the basic assumptions for the function h and (6), we can easily obtain that

$$|A_R(k,\theta',\theta)| \leq \int_{\mathbb{R}^2} |h(y,|u|)u(y,k,\theta) - h_0(y)u_0|dy \leq C\int_{\mathbb{R}^2} (\alpha(y)+\beta(y))|u_{sc}(y)|dy$$
$$\leq C\|u_{sc}\|_\infty \left(\|\alpha\|_1 + \|\beta\|_1\right) \to 0, \qquad |k| \to \infty.$$

We have used here the fact that the basic assumptions for the function h guarantee that the functions α and β both belong to $L^1(\mathbb{R}^2)$.

Hence, for k large, we have approximately that

$$A(k,\theta',\theta) \approx F(h_0)\big(k(\theta-\theta')\big).$$

These considerations and real valuedness of h suggest and justify the following definition: We define the inverse Born approximation q_B via the equality

$$A(k,\theta',\theta) = F(q_B)(k(\theta'-\theta))$$

which is understood in the sense of tempered distributions. In order to recover main singularities of h_0 from q_B, we must study their difference and show that it is locally less singular than h_0. To this end, we have the following main result from [5].

Theorem 2. Let σ be as in (7). Then

$$q_B - h_0 \in H^t_{loc}(\mathbb{R}^2),$$

where $t < 3 - 4/p$ if $4/3 < p \leq 3/2$ and $t < 1 - 1/p$ if $3/2 < p \leq \infty$.

Remark 2. Theorem 2 means that, for $4/3 < p < \infty$, the main singularities of h_0 can be recovered from the inverse scattering Born approximation q_B with full data D. In the case of $p = \infty$, we have no singularities but may have finite jumps. Under such circumstances, we record the following special case.

Corollary 3. If a piecewise smooth compactly supported function h_0 contains a jump over a smooth curve, then the curve and the height function of the jump are uniquely determined by the full scattering data. Especially, for the function h_0 being the characteristic function of a smooth bounded domain, this domain is uniquely determined by the full scattering data.

To conclude, in this section of the study, the distinctiveness of the direct problem for the nonlinearities h that meet the necessary features was demonstrated. These qualities permit the existence of local singularities and do not necessitate compact support, but rather demand a certain level of decay at infinity. By examining comparable characteristics, we were able to determine the asymptotic behaviour of scattering solutions. This allows us to obtain the scattering data necessary for investigating inverse scattering problems. It is important to mention that both results were demonstrated without assuming that the norm of the nonlinearity is minimal, which is a requirement in dimensions three and higher.

What can we regard as the main result of this section is the Saito's formula since it implies a uniqueness result and a representation formula for our unknown function h_0. In addition, we managed to extract more information about the nonlinearity by applying the method of Born approximation. More precisely, the main singularities (or jumps over smooth curve) of h_0 can be recovered from the Born approximation q_B which corresponds to the full scattering data D.

2.2. Backscattering and fixed angle data

In this section, we consider backscattering data D_B and fixed angle scattering data D_A following [12]. Using (10) we introduce the inverse backscattering and inverse fixed angle scattering Born approximations q_B^b and $q_B^{\theta_0}$ as follows:

$$A(k, -\theta, \theta) = F(q_B^b)(2k\theta)$$

and

$$A(k.\theta'.\theta_0) = F(q_B^{\theta_0})(k(\theta' - \theta_0)).$$

where θ_0 is fixed.

Furthermore, we assume in addition that the nonlinearity h possesses the Taylor expansion:

$$h(x, 1+s) = h(x,1) + \partial_s h(x,1)s + \frac{1}{2}\partial_s^2 h(x, s^*)s^2, \qquad 1 < s^* < 1+s$$

where

$$|\partial_s h(x,1)| \leq \eta_1(x), \qquad |\partial_s^2 h(x, s^*)| \leq \eta_2(x)$$

uniformly in $s \in (0, s_0), s_0 > 0$ and with $\eta_1, \eta_2 \in L_\sigma^p(R^2)$, where σ as in (7). From this we obtain the expansion:

$$h(x,|u|)u = h_0(x)u_0 + g_1(x)u_{sc} + g_2(x)u_0^2\overline{u_{sc}} + \eta(x)O(|u_{sc}|^2),$$

where $g_2(x) = \frac{1}{2}\partial_s h(x, 1)$, $g_1(x) = h_0(x) + g_2(x)$ and $\eta \in L_\sigma^p(\mathbb{R}^2)$ with the same σ as above.

Again, the main result for the recovery of main singularities is formulated as the following theorem.

Theorem 3. Let σ be as in (7) with $2 < p \leq \infty$. Suppose in addition

$$\widehat{h_0}, \widehat{g_2} \in W_s^1(\mathbb{R}^2), \qquad 1 < s < 2,$$

and $2 < s' < p \leq \infty$, where s' is the Hölder conjugate of s. Then

$$q_B^{\theta_0} - h_0, q_B^b - h_0 \in H_{loc}^t(\mathbb{R}^2)$$

for any $t < 1 - 1/p$ if $1 < s \leq 4/3$ and for any $t < \min\{1 - 1/p, 4/s - 2\}$ if $4/3 < s < 2$.

Let us sketch the main ideas in the proof of Theorem 3. Using the definition, we may expand the difference in several terms as

$$q_B^{\theta_0} - h_0 = q_1^{\theta_0} + q_\infty^{\theta_0} + q_2^{\theta_0} + q_R^{\theta_0}.$$

In straightforward manner one sees that $q_\infty^{\theta_0} \in C^\infty(\mathbb{R}^2)$ and $q_2^{\theta_0}$, $q_R^{\theta_0} \in H^t(\mathbb{R}^2)$ with $t < 3 - 4/p$ if $4/3 < p \leq 3/2$ and $t < 1 - 1/p$ if $3/2 < p \leq \infty$. The first nonlinear term $q_1^{\theta_0}$ cannot be analyzed directly from its definition. Instead, we first proved the representation

$$q_1^{\theta_0}(x) = -F_4^{-1}\left(pv\,\frac{\widehat{g_1}(\xi)\,\widehat{h_0}(\eta)(\xi+\eta,\theta_0)}{(|\eta|^2(\xi+\eta)-|\xi+\eta|^2\eta,\theta_0)}\right)(x,x)$$
$$-F_4^{-1}\left(pv\,\frac{\widehat{g_2}(\xi)\,\widehat{h_0}(\eta)(\xi+\eta,\theta_0)}{(|\eta|^2(\xi+\eta)+|\xi+\eta|^2\eta,\theta_0)}\right)(x,x),$$

where F_4^{-1} denotes the four-dimensional inverse Fourier transform. This formula might have independent interest too, but primarily it allows us to prove the following regularity: the term $q_1^{\theta_0}$ belongs to the space

1. $C(\mathbb{R}^2)\cap L^{\infty}(\mathbb{R}^2)$ if $1 < s < 4/3$;
2. $H^1(\mathbb{R}^2)$ if $s = 4/3$;
3. $H^t(\mathbb{R}^2)$, $t<4/s-2$ if $4/3 < s < 2$.

If we combine all these steps, we obtain Theorem 3 for fixed angle scattering. The inverse backscattering Born approximation is treated similarly. Namely, we write

$$q_B^b - h_0 = q_1^b + q_\infty^b + q_2^b + q_R^b.$$

The latter four terms have exactly the same regularity results as their counterparts in fixed angle scattering. For the first nonlinear term, the representation is now

$$q_1^b(x) = 4F_4^{-1}\left(pv\,\frac{\widehat{g_1}(\xi)\,\widehat{h_0}(\eta)}{(\xi,\eta)}\right)(x,x) - 4F_4^{-1}\left(pv\,\frac{\widehat{g_2}(\xi)\,\widehat{h_0}(\eta)}{(\xi+2\eta,\eta)}\right)(x,x).$$

The additional assumption $\widehat{h_0}\in W_s^1(\mathbb{R}^2)$ in Theorem 3 implies that $h_0\in L_1^{s'}(\mathbb{R}^2)$. This explains why we restrict $s' < p$. By Sobolev embedding, we obtain

$$q_B^{\theta_0} - h_0,\, q_B^b - h_0 \in W_p^{\epsilon}(\mathbb{R}^2)$$

with some positive $\epsilon < \min\{1/p, 1 - 2/p\}$. Hence, h_0 is locally more singular than either of these differences. That's why both Born approximations recover the main singularities of h_0. On the other hand, we may perform the comparison also in the scale of Sobolev spaces. Indeed, if $h_0\in H^r_{comp}(\mathbb{R}^2)$ with some $0 < r < 1$, then

$$q_B^{\theta_0} - h_0,\, q_B^b - h_0 \in H^t_{loc}(\mathbb{R}^2)$$

for any $t < 2r$ if $0 < r \le 1/3$ and for any $t < (1 + r)/2$ if $1/3 < r < 1$. In both cases this t can be made bigger than r. It means that we can reconstruct all singularities from Sobolev space H^r, $0 < r < 1$ from data D_B and D_A by the method of Born approximation.

Corollary 4. If a piecewise smooth compactly supported function h_0 contains a jump over a smooth curve, then the curve and the height function of the jump are uniquely determined by backscattering and fixed angle scattering data. Especially, for the function h_0 being the characteristic function of a smooth bounded domain, this domain is uniquely determined by backscattering and fixed angle scattering data.

Concluding, in this section we proved that all singularities and jumps (in the absence of a uniqueness theorem) of the nonlinearity h can be recovered from the inverse scattering Born approximation corresponding to fixed angle scattering and backscattering data D_A and D_B, respectively. No assumptions about the smallness of the norm of nonlinearity h were used as it were in previous publications even in the linear case.

2.3. Fixed energy data

The two-dimensional fixed energy problemx1 was a long-standing open problem. In the case of linear Schrödinger operator, the first uniqueness and reconstruction algorithm was proved by Nachman [13] via $\partial^{-}\partial^{-}$-methods for potentials of conductivity type. Sun and Uhlmann [14] proved uniqueness for potentials satisfying nearness conditions to each other. The question of global uniqueness for the linear Schrödinger equation with fixed energy was settled only in 2008 by Bukhgeim [15] for compactly supported potentials from $L^p, p > 2$. This result has recently been improved and extended to related inverse problems (see for example [16] and [17]). Note that Grinevich and Novikov [18] proved that in two dimensions there are nonzero real potentials of the Schwartz class with zero scattering amplitude at fixed positive energy. Thus, the compactness of the supports of the potentials is very natural condition in our considerations.

The results of this section are proved in [19] and they slightly generalize the linear case to a special type of nonlinearity. It turned out that (as we can see in this section) inverse fixed energy scattering problem is much more difficult than the others. In fixed energy scattering problem, instead of the scattering solutions (4) we need the so-called complex geometrical optics solutions. Complex geometrical optics (CGO) solutions or exponentially growing solutions of the form:

$$u(x,z) = e^{i(x,z)}\big(1 + R(x,z)\big), \quad z \in \mathbb{C}^2, (z,z) = 0$$

with R decaying at infinity for $|z|$ large for the homogeneous Schrödinger equation

$$-\Delta u(x) + h(x, |u(x)|)u(x) = 0$$

are obtained as follows. Substituting (14) into (15) yields

$$-\Delta R - 2i(z,\nabla)R + h\big(x, \big|e^{i(x,z)}(1+R)\big|\big)(1+R) = 0,$$

It means that using the Faddeev Green's function

$$g_z(x) = \frac{1}{4\pi^2}\int_{\mathbb{R}^2} \frac{e^{-i(x,\xi)}}{\xi^2 + 2(z,\xi)} d\xi$$

as the fundamental solution of the differential operator

$$-\Delta - 2i(z, \nabla)$$

we see that the function R solves the integral equation

$$R(x,z) = -\int_{\mathbb{R}^2} g_z(x-y)h(y, |e^{i(y,z)}(1+R(y,z))|)(1+R(y,z))dy.$$

It remains to establish unique solvability of this equation. To this end we again use iterations in the sense of next theorem.

We assume that h is compactly supported in Ω⊂R2Ω⊂ℝ2 and

- $|h(x,s)| \leq \alpha(x)$ with some $\alpha \in L^2(\Omega)$ and $s \geq 0$
- $|h(x, |e^{i(x,z)}(1+R_1)|) - h(x, |e^{i(x,z)}(1+R_2)|)| \leq \beta(x)|R_1 - R_2|$ with some $\beta \in L^2(\Omega)$ and for any $R_1, R_2 \in L^{\infty}\mathbb{R}^2$ such that $||R_j||_\infty < 1 - \delta, j = 1, 2$ for some fixed $\delta \in (0, 1)$ and for any $z \in \mathbb{C}^2$
- $||\alpha||_2, ||\beta||_2 > 0$

Theorem 4. Under the above conditions for h, there exists a constant $C_0 > 0$ such that for all $|z| \geq C_0$ the equation (16) has a unique solution in $L^{\infty}\mathbb{R}^2$ and this solution can be obtained as $\lim_{j\to\infty} R_j$ in $L^{\infty}(\mathbb{R}^2)$, where

$$R_{j+1}(x,z) = -\int_{\mathbb{R}^2} g_z(x-y)h\left(y, \left|e^{i(y,z)}\left(1+R_j(y,z)\right)\right|\right)\left(1+R_j(y,z)\right)dy, j = 0,1,\ldots$$

with $R_0 = 0$. Moreover, the following estimates hold

$$||R||_\infty \leq \frac{c}{|z|^\gamma}, \left\|R - R_j\right\|_\infty \leq \frac{c_j}{|z|^{\gamma(j+1)}}, \; j = 0,1,\ldots.$$

The proof of Theorem 4 is based on the fact that for any $\gamma < 1$ there is constant $c_\gamma > 0$ such that

$$||g_z * f||_\infty \leq \frac{c_\gamma}{|z|^\gamma}||f||_2, \; |z| > 1,$$

for any $f \in L^2(\Omega)$, see [20].

Turning now to the inverse fixed energy scattering problem, we define the scattering transform by

$$T_h(\xi) = \int_{\mathbb{R}^2} e^{i(x,\xi)} h(x, e_0|1+R(x,z)|)(1+R(x,z))dx, \qquad |\xi| \geq \sqrt{2}\, C_0$$

and $T_h(\xi) = 0$ for $|\xi| < \sqrt{2}C_0$. Here $z = \frac{1}{2}(\xi - iJ\xi)$,

$$J = \begin{pmatrix} 0 & 1 \\ -1 & 0 \end{pmatrix}$$

and $e_0 = |e^{i(x,z)}| = e^{\frac{1}{2}(x_1\xi_2 - x_2\xi_1)}$. What is more, we have the uniform limit $T_h(\xi) = \lim_{j\to\infty} T_{h,j}(\xi),$, where

$$T_{h,j}(\xi) = \int_{\mathbb{R}^2} e^{i(x,\xi)} h(x, e_0|1 + R_j(x,z)|)\left(1 + R_j(x,z)\right) dx.$$

We point out that the scattering transform is somehow an auxiliary object (see D_E in Introduction). But it is connected to the scattering amplitude as follows. It is well known that the scattering amplitude at fixed energy uniquely determines the Dirichlet-to-Neumann map $\Lambda_{h-k_0^2}$ which in turn uniquely determines the scattering transform (see the details, for example, in [19] and [20]). Recall that $\Lambda_h f = \partial_\nu u$, where u satisfies the Dirichlet problem:

$$\begin{cases} -\Delta u + h(x, |u|)u = 0, x \in \Omega \\ \quad u(x) = f(x), x \in \partial\Omega. \end{cases}$$

Here, Ω is a domain with boundary $\partial\Omega$ and outward unit normal vector ν.

Next, we define the inverse fixed energy Born approximation by

$$q_B^f(x) = F^{-1}\left(T_h(\xi)\right)(x)$$

In contrast to the preceding inverse problems, we now set the unknown function to be

$$h_0(x) = F^{-1}\left(T_{h,0}(\xi)\right)(x).$$

In linear case h_0 is the actual potential appearing in the Schrödinger equation, but otherwise the connection to physical scatterers is not known to us.

We assume that the nonlinearity h admits the Taylor expansion

$$h(x, e_0(1+s)) = h(x, e_0) + \partial_s h(x, e_0(1+s))\big|_{s=0} s + O(\beta_1(x)s^2),$$

where $|\partial_s h(x, e_0(1+s))|_{s=0}| \leq \beta_1(x)$ and $O(\beta_1(x)s^2)$ with $\beta_1(x) \in L^2(\Omega)$ and with small s in the neighbourhood of zero and where O is uniform in $x \in \Omega$ and such s.

Suppose in addition that the nonlinearity h satisfies the asymptotic expansions

$$h(x, e_0) = \sum_{j=0}^{\infty} \frac{\alpha_j(x)}{|\xi|^j}, \quad h(x, e_0|1 + R_1|) = \sum_{j=0}^{\infty} \frac{\tilde{\alpha}_j(x)}{|\xi|^j}$$

Where. Then we have the following main result concerning the recovery of singularities of $h_0(x)$ defined by (18).

Theorem 5. Under the foregoing conditions for the potential function h, it is true that

$$q_B^f - h_0 \in H^t(\mathbb{R}^2)$$

for any $t < 1$ modulo $C^{\infty}\mathbb{R}^2$ - functions.

Remark 3. The embedding theorem for Sobolev spaces says that the difference $q_B^f - h_0$ belongs to $L^q(R^2)$ for any $q < \infty$ modulo $C^{\infty}(R^2)$ functions. It means that all singularities from $L^p_{loc}(R^2)$, $p<\infty$ of unknown function h_0 can be obtained exactly by the Born approximation which corresponds to the inverse scattering problem with fixed positive energy.

We note that under fixed energy data we have some additional assumptions on h. This limits the applicability of the main result to, for example, saturation type nonlinearities. In particular, cubic nonlinearity is excluded from these considerations. Moreover, the unknown function h_0 has no direct connection to original scatterers in nonlinear cases.

3. NUMERICAL EXAMPLES

Here we discuss the numerical computation of the Born approximation first proposed in [21] for backscattering and fixed angle data. We assume that the scatterer $h_0(x)$ is supported in the rectangle $R_0=[-1,1]\times[-1,1]\subset\mathbb{R}^2$. We consider the following examples:

Example 1: $h(x,|u|) = \chi_\Omega(x)|u|$ (linear)

Example 2: $$h(x,|u|) = \chi_\Omega(x)|u|\frac{|u|^2}{1+|u|^2}$$ (saturation)

Example 3: $$h(x,|u|) = 0.75\chi_{\Omega_1}(x)|u| + \chi_{\Omega_2}(x)\frac{|u|^2}{1+|u|^2}$$ (linear, saturation)

Example 4: $$h(x,|u|) = 0.75\chi_{\Omega_2}(x)\frac{|u|^2}{1+|u|^2} + \chi_{\Omega_1}(x)|u|$$ (saturation, linear)

Here $\Omega, \Omega_1, \Omega_2$ are the shifted ellipses rotated in angle Θ theta (counter-clockwise) detailed in Table 1 and shown in Figure 1.

Table 1. Geometries of the ellipses Ω, Ω_1 and Ω_2

	Semi axis	Θ	Centre
Ω	1/2, 1/5	$17\pi/18$	(-0.3, - 0.4)
Ω_1	1/2, 1/4	$\pi/3$	(0.5, 0)
Ω_2	1/3, 1/5	$2\pi/3$	(-0.5, 0)

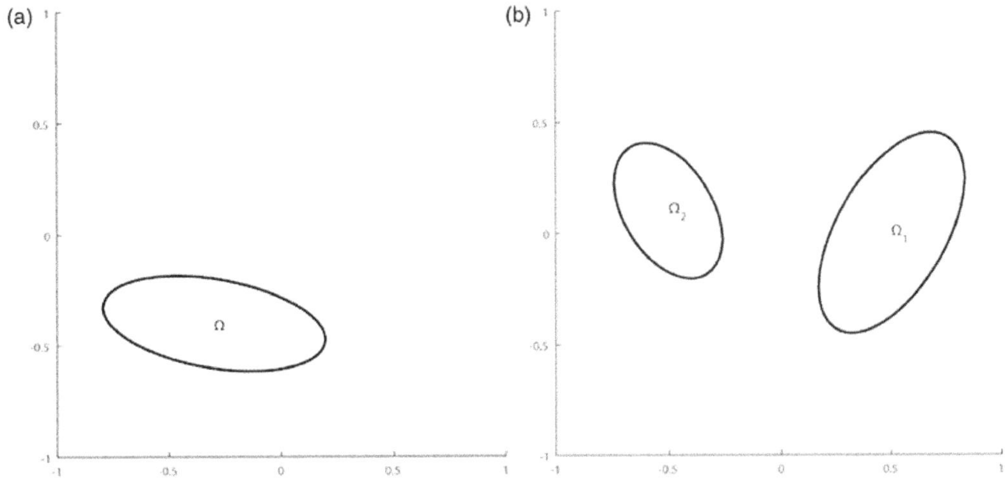

Figure 1. Geometries of the ellipses Ω, Ω_1 and Ω_2.

Consider the Born approximation for full data (11) in the form

$$\int_{\mathbb{R}^2} e^{ik(\theta'-\theta,y)} f(y)dy = A(k,\theta',\theta), \ \ f = q_B.$$

To discretize the unknown function f we divide the rectangle R_0 into $N = n \times n$ disjoint subrectangles r_j of equal size, i.e.

$$R_0 = \bigcup_{j=1}^{N} r_j.$$

Then, we represent f on R_0 in piecewise constant form:

$$f(y) = \sum_{j=1}^{N} f_j \chi_{r_j}(y),$$

where f_j's are the unknown values on r_j's. Substitution into (19) yields

$$\sum_{j=1}^{N} f_j \int_{r_j} e^{ik(\theta'-\theta,y)} dy = A(k,\theta',\theta).$$

Evaluating this at some points $k = k_s$, $\theta' = \theta_t'$ and $\theta = \theta_p$ leads us to

$$\sum_{j=1}^{N} f_j \int_{r_j} e^{ik_s(\theta_t'-\theta_p,y)} dy = A(k_s,\theta_t',\theta_p), \qquad s = 1,\dots,N_1, t = 1,\dots,N_2, p = 1,\dots,N_3.$$

If we denote $M = N_1N_2N_3$, we may form one linearized index $l = l(s,t,p), l = 1, \dots, M$ and write the latter equation as

$$\sum_{j=1}^{N} f_j E(j,l) = g_l, \ \ l = 1,\dots,M,$$

where $E(j,l)=\int_{\Gamma_j} e^{ik_s(\theta_t'-\theta_p,y)}\,dy$ are easily evaluated and $g_l = A(k_s, \theta_t', \theta_p)$ needs to be computed. The computation of g_l is carried out using numerical integration, see [21] for details. In matrix form (20) is clearly $Ef = g$.

For backscattering data and fixed angle data the system (20) is modified accordingly. We note that the system $Ef = g$ does not depend on the scatterer but only on data type and measurement setup.

The fixed energy case differs from the first three data sets. In fixed energy inversion we approximate the scattering transform as

$$T_h(\xi) \approx T_{h,1}(\xi).$$

We choose $M = m \times m$ points ξ uniformly from the rectangle $[-s, s] \times [-s, s]$. The function $T_{h,1}(\xi)$ is evaluated by numerical integration, see [19,21] for details. Then the inverse Born approximation (17) is computed similarly to (19).

We use the following parameter values:

$$k_s = s\,, \theta_p = \theta_p' = \left(\cos\frac{p2\pi(N_2-1)}{N_2}, \sin\frac{p2\pi(N_2-1)}{N_2}\right)$$

We use $N = 2500$ and $N_1 = 12$ for each data set. For full data we use $N_2 = N_3 = 6$. For backscattering and fixed angle data, we use $N_2 = 24$. For fixed energy scattering we use $m = 40$ and $s = 6$.

In all four cases we obtain the linear system $Ef = g$ whose coefficient matrix E is of size $M \times N$. The data g is corrupted with zero mean Gaussian noise with standard deviation $\sigma = 0.01 \max |g|$.

The size M as well as the ranks $r(E)$ and (approximate) condition numbers $\log_{10}\kappa(E)$ measuring the ill-posedness of the linear system $EF = g$ are shown in Table 2.

Table 2.Matrix sizes, ranks and condition numbers

Data	M	r(E)	$\log_{10}$(E)
Full	1728	360	63
Backscattering	288	281	15
Fixed angle	288	241	17
Fixed energy	1600	268	15

As the linear system is rank-deficient and ill-posed, we use regularization method to solve it. More precisely, we apply the total variation method (TV) which is defined as

$$f = arg\min_z\{||Ez-g||_2^2 + \delta||Lz||_1\},$$

where the matrix L implements the differences between neighbouring elements in horizontal and vertical directions (for details, see [21]). As in [21] we formulate this minimization problem as a quadratic problem in standard form for more efficient solution. As the regularization parameter we use $\delta = 2 \cdot 10^{-3}$ for D_E and $\delta = 10^{-3}$ otherwise.

All computations are carried out in a UNIX system with 512 GB of memory and 40 logical CPU cores, each running at 2.8 GHz. The software platform is MATLAB R2015b. We have used 12 workers in parallel in computing the right-hand side g. We note that a desktop PC with 8 logical cores running at 3.4 GHz and 16 GB of memory is also capable of our computations, but with 7 workers it is considerably slower in computing g.

The table 3 displays the computational durations for both the creation and solution of the linear system. It should be noted that the right-hand side g includes artificially generated data, and the process of obtaining actual physical measurements may vary in duration.

Table 3. Computational times

		Example 1		Example 3	
Data	E	g	TV	g	TV
Full	1.3s	47s	70s	3m20s	63s
Backscattering	0.4s	9s	59s	37s	45s
Fixed angle	0.3s	9s	66s	33s	58s
Fixed energy	0.1s	4m11s	78s	-	-

The contour plots of scatterers $h_0(x)$ and their TV reconstructions via Born approximation from full data, backscattering data and fixed angle data (with $\theta_0 = (1, 0)$) are shown in Figures 2–5 for all examples. For fixed energy scattering we only consider the linear Example 1, since otherwise we do not have direct comparison to a scatterer. The TV reconstruction is shown in Figure 6. In each figure solid white line indicates the true geometry of the scatterer.

We see that the location of the scatterer is located quite nicely in all cases. The shape of the scatterer is best seen from full data and backscattering data. By computing the Born approximation from full angle data, we close an open problem from [5].

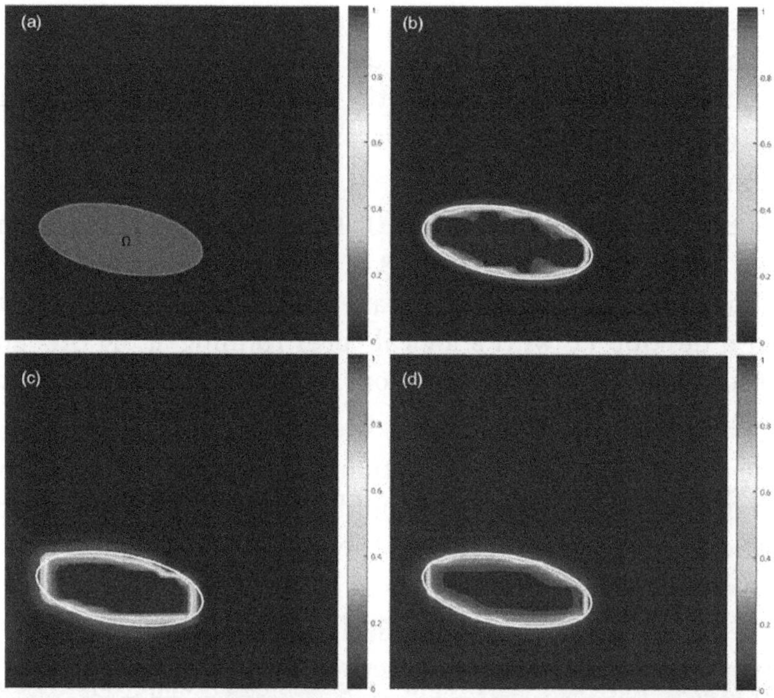

Figure 2. Scatterer $h_0(x)$ and its TV reconstruction via Born approximation, Example 1. (a) scatterer; (b) full data; (c) backscattering data; (d) fixed angle data.

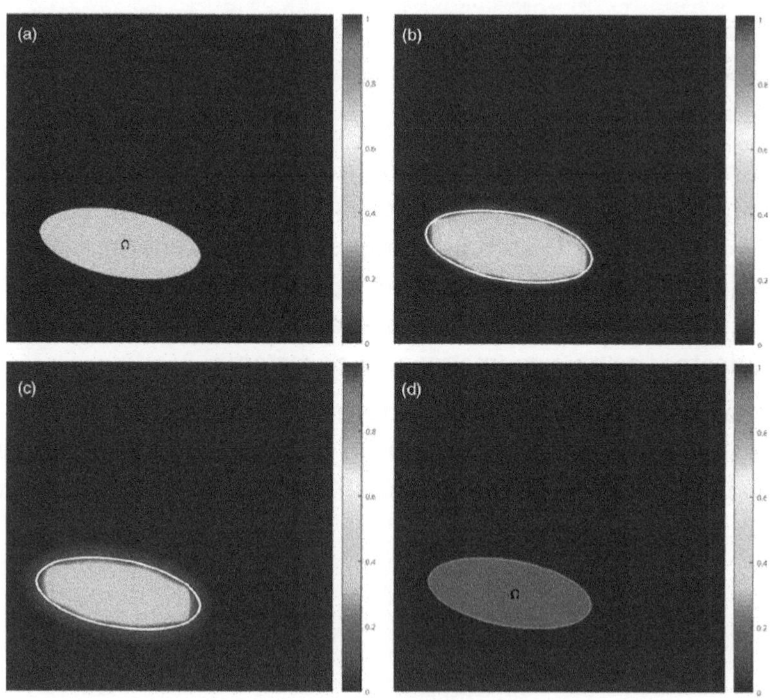

Figure 3. Scatterer $h_0(x)$ and its TV reconstruction via Born approximation, Example 2. (a) scatterer; (b) full data; (c) backscattering data; (d) fixed angle data

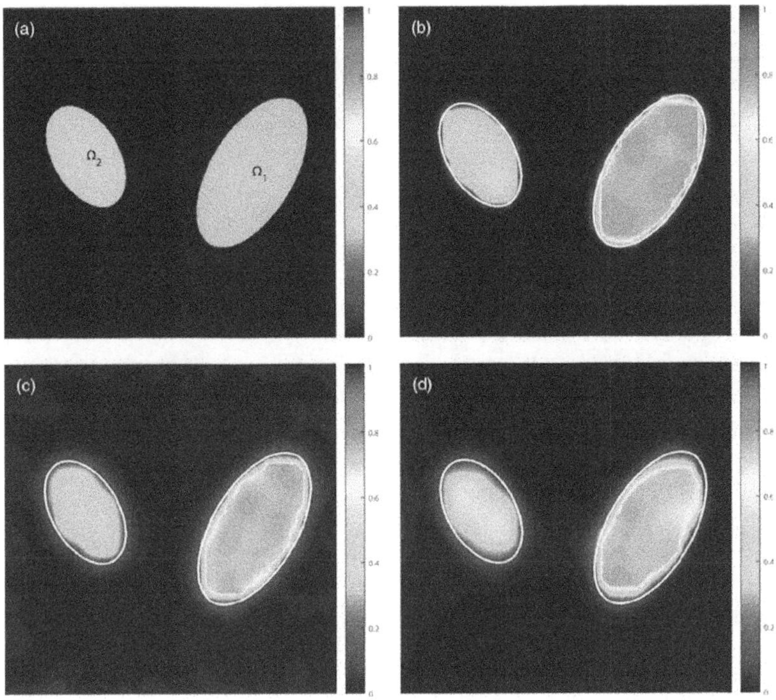

Figure 4. Scatterer $h_0(x)$ and its TV reconstruction via Born approximation, Example 3. (a) scatterer; (b) full data; (c) backscattering data; (d) fixed angle data.

Figure 5. Scatterer $h_0(x)$ and its TV reconstruction via Born approximation, Example 4. (a) scatterer; (b) full data; (c) backscattering data; (d) fixed angle data.

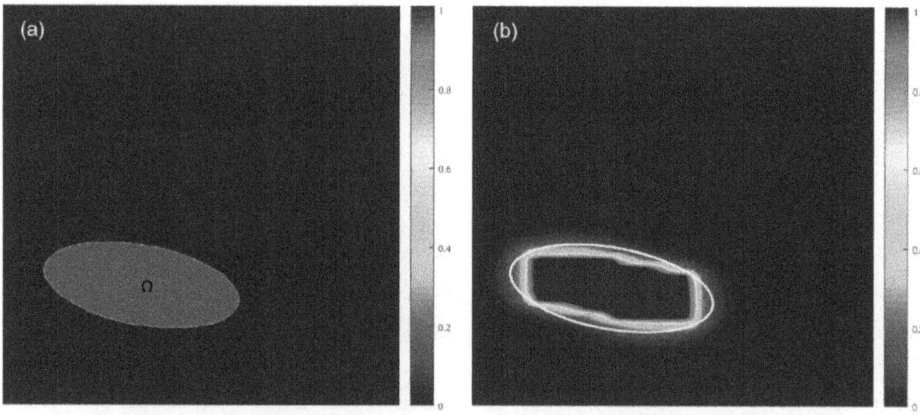

Figure 6. Scatterer $h_0(x)$ and its TV reconstruction via Born approximation, Example 1. (a) scatterer; (b) fixed energy.

REFERENCES

1. Bang O, Krolikowski W, Wyller J, Rasmussen J J. Collapse arrest and soliton stabilization in nonlocal nonlinear media. Phys. Rev. E. 2002;66:046619.
2. Berge L, Gouedard C, Schjodt-Eriksen J, Ward H. Filamentation patterns in Kerr media versus beam shape robustness, nonlinear saturation and polarization state. Physica D. 2003;176:181–211.
3. Dreischuh A, Paulus G G, Zacher F, Grasbon F, Neshev D, Walther H. Modulational instability of multi-charged optical vortex solitons under saturation on the nonlinearity. Phys. Rev. E. 1999;60:7518–7522.
4. Kartashov Y V, Vysloukh V A, Torner L. Two-dimensional cnoidal waves in Kerr-type saturable nonlinear media. Phys. Rev. E. 2003;68:015603.
5. Serov V, Harju M, Fotopoulos G. Direct and inverse scattering for nonlinear Schrödinger equation in 2D. J. Math. Phys.. 2012;53:123522.
6. Serov V, Sandhu J. Inverse backscattering problem for the generalized nonlinear Schrödinger operator in two dimensions. J. Phys. A: Math. Theor.. 2010;43:325206.
7. Serov V. An inverse Born approximation for the general nonlinear Schrödinger operator on the line. J. Phys. A: Math. Theor.. 2009;42:332002.
8. Harju M. On the direct and inverse scattering problems for a non-linear three-dimensional Schrödinger equation [dissertation]. University of Oulu: 2010.
9. Harju M, Serov V. Three-dimensional direct and inverse scattering for the Schrödinger equation with a general nonlinearity. Operator Theory: Advances and Applications. 2014;236:257–273.

10. Isakov V, Sylvester J. Global uniqueness for a semi-linear elliptic inverse problem. Commun. Pure Appl. Math.. 1994;47:1403–1410.

11. Lechleiter A. Explicit characterization of the support of non-linear inclusions. Inverse Probl. Imaging. 2011;5:675–694.

12. Fotopoulos G, Harju M, Serov V. Inverse fixed angle scattering and backscattering for a nonlinear Schrödinger equation in 2D. Inv. Probl. Imag.. 2013;7:183–197.

13. Nachman A I. Global uniqueness for a two-dimensional boundary value problem. Ann. Math.. 1996;143:71–96.

14. Sun Z, Uhlmann G. Recovery of singularities for formally determined inverse problems. Commun. Math. Phys.. 1993;153:431–445.

15. Bukhgeim A L. Recovering a potential from Cauchy data in the two-dimensional case. J. Inverse Ill-Posed Problems. 2008;16:19–33.

16. Blåsten E. On the Gel'Fand–Calderón inverse problem in two dimensions [dissertation]. University of Helsinki:2013.

17. Imanuvilov OY, Yamamoto M. Inverse boundary value problem for Schrödinger equation in two dimensions. SIAM J. Math. Anal.. 2012;44:1333–1339.

18. Grinevich P, Novikov R. Transparent potentials at fixed energy in two dimensions. Fixed-energy dispersion relations for the fast decaying potentials. Commun. Math. Phys.. 1995;174:409–446.

19. Fotopoulos G, Serov V. Inverse fixed energy scattering problem for the two-dimensional nonlinear Schrödinger operator. Inv. Probl. Sci. Eng.. 2015; DOI: 10.1080/17415977.2015.1055263

20. Serov V. Inverse fixed energy scattering problem for the generalized nonlinear Schrödinger operator. Inv. Probl.. 2012;28:025002.

21. Harju M. Numerical computation of the inverse Born approximation for the nonlinear Schrödinger equation in two dimensions. Comp. Meth. Appl. Math.. 2016;16:133–143.

06 APPLICATION OF LIKELIHOOD RATIO TESTS IN MULTIVARIATE LINEAR MODELS

1. INTRODUCTION: AN OVERVIEW

In this chapter, we review statistical inference, especially likelihood ratio criterion (LRC) in multivariate linear model, focusing on matrix theory. Consider a multivariate linear model with p response variables $y_1, ..., y_p$ and k explanatory or dummy variables $x_1, ..., x_k$. Suppose that $\mathbf{y} = (y_1, ..., y_p)'$ and $\mathbf{x} = (x_1, ..., x_k)'$ are measured for n subjects, and let the observation of the ith subject be denoted by $\mathbf{y}_i$ and $\mathbf{x}_i$. Then, we have the observation matrices given by

$$\mathsf{Y} = (y_1, y_2, \ldots, y_n)', \quad \mathrm{X} = (x_1, x_2, \ldots, x_n)'.$$

It is assumed that $\mathbf{y}_1, ..., \mathbf{y}_n$ are independent and have the same covariance matrix Σ. We express the mean of $\mathbf{Y}$ as follows:

$$\mathrm{E}(\mathbf{Y}) = \boldsymbol{\eta} = (\boldsymbol{\eta}_1, \ldots, \boldsymbol{\eta}_p).$$

A multivariate linear model is defined by requiring that

$$\eta_i \in \Omega \text{ for all } i = 1, \ldots, p,$$

where Ω is a given subspace in the n dimensional Euclid space R^n. A typical Ω is given by

$$\Omega = \mathcal{R}[\mathbf{X}] = \{\boldsymbol{\eta} = \mathbf{X}\boldsymbol{\theta}: \boldsymbol{\theta} = (\theta_1, \ldots, \theta_k)', -\infty < \theta_i < \infty, i = 1, \ldots, k\}$$

Here, $\mathcal{R}[\mathbf{X}]$ is the space spanned by the column vectors of $\mathbf{X}$. A general theory for statistical inference on the regression parameter ΘΘ can be seen in texts on multivariate analysis, e.g., see [1–8]. In this chapter, we discuss with algebraic approach in multivariate linear model.

In Section 2, we consider a multivariate regression model in which $x_i's$ are explanatory variables and $\Omega = \mathcal{R}[\mathbf{X}]$. The maximum likelihood estimator (MLE)s and likelihood ratio criterion (LRC) for Θ2=OΘ2=O are derived by using projection matrices. Here, Θ=(Θ1Θ2).Θ=Θ1Θ2. The distribution of LRC is discussed by multivariate Cochran theorem. It is pointed out that projection matrices play an important role. In Section 3, we give a summary of projection matrices. In Section 4, we consider to test an additional information hypothesis of $\mathbf{y}_2$ in the presence of $\mathbf{y}_1$, where $\mathbf{y}_1 = (y_1, ..., y_q)'$ and $\mathbf{y}_2 = (y_{q+1}, ..., y_p)'$. In Section 5, we consider testing problems in discriminant analysis. Section 6 deals with a generalized multivariate linear model which is also called the growth curve model. Some related problems are discussed in Section 7.

2. MULTIVARIATE REGRESSION MODEL

In this section, we consider a multivariate regression model on p response variables and k explanatory variables denoted by $\mathbf{y} = (y_1, ..., y_p)'$ and $\mathbf{x} = (x_1, ..., x_k)'$, respectively. Suppose that we have the observation matrices given by (1.1). A multivariate regression model is given by

$$\mathbf{Y} = \mathbf{X}\boldsymbol{\Theta} + \mathbf{E},$$

where Θ is a $k \times p$ unknown parameter matrix. It is assumed that the rows of the error matrix $\mathbf{E}$ are independently distributed as a p variate normal distribution with mean zero and unknown covariance matrix Σ, i.e., $N_p(0, \Sigma)$.

Let $L(\Theta, \Sigma)$ be the density function or the likelihood function. Then, we have

$$-2\log L(\boldsymbol{\Theta}, \boldsymbol{\Sigma}) = n\log|\boldsymbol{\Sigma}| + \mathrm{tr}\boldsymbol{\Sigma}^{-1}(\mathbf{Y} - \mathbf{X}\boldsymbol{\Theta})'(\mathbf{Y} - \mathbf{X}\boldsymbol{\Theta}) + np\log(2\pi).$$

The maximum likelihood estimators (MLE) $\hat{\Theta}$ and $\hat{\Sigma}$ of Θ and Σ are defined by the maximizers of $L(\Theta, \Sigma)$ or equivalently the minimizers of $-2\log L(\Theta, \Sigma)$.

Theorem 2.1 *Suppose that* $\mathbf{Y}$ *follows the multivariate regression model in* (2.1). *Then, the* MLEs *of* $\Theta\Theta$ *and* $\Sigma\Sigma$ *are given as*

$$\hat{\boldsymbol{\Theta}} = (\mathbf{X}'\mathbf{X})^{-1}\mathbf{X}'\mathbf{Y},$$

$$\hat{\boldsymbol{\Sigma}} = \frac{1}{n}(\mathbf{Y} - \mathbf{X}\hat{\boldsymbol{\Theta}})'(\mathbf{Y} - \mathbf{X}\hat{\boldsymbol{\Theta}}) = \frac{1}{n}\mathbf{Y}'(\mathbf{I}_n - \mathbf{P}_\mathbf{X})\mathbf{Y},$$

where $\mathbf{P}_X = \mathbf{X}(\mathbf{X}'\mathbf{X})^{-1}\mathbf{X}'$. Further, it holds that

$$-2\log L(\hat{\boldsymbol{\Theta}}, \hat{\boldsymbol{\Sigma}}) = n\log|\hat{\boldsymbol{\Sigma}}| + np\{\log(2\pi) + 1\}.$$

Theorem 2.1 can be shown by a linear algebraic method, which is discussed in the next section. Note that $\mathbf{P}_X$ is the projection matrix on the range space Ω=R[X]. It is symmetric and idempotent, i.e.

$$\mathbf{P}_\mathbf{X}' = \mathbf{P}_\mathbf{X}, \quad \mathbf{P}_\mathbf{X}^2 = \mathbf{P}_\mathbf{X}.$$

Next, we consider to test the hypothesis

$$H : \mathrm{E}(\mathbf{Y}) = \mathbf{X}_1\boldsymbol{\Theta}_1 \quad \Leftrightarrow \quad \boldsymbol{\Theta}_2 = \mathbf{O},$$

against K: $\Theta_2 \neq \mathrm{O}$, where $\mathbf{X} = (\mathbf{X}_1\ \mathbf{X}_2)$, $\mathbf{X}_1$; $n \times j$ and $\Theta = (\Theta_1'\ \Theta_2')$, Θ_1; $j \times p$. The hypothesis means that the last $k - j$ dimensional variate $\mathbf{x}_2 = (x_{j+1}, ..., x_k)'$ has no additional information in the presence of the first j variate $\mathbf{x}_1 = (x_1, \ldots, x_j)'$. In general, the likelihood ratio criterion (LRC) is defined by

$$\lambda = \frac{\max_H L(\boldsymbol{\Theta}, \boldsymbol{\Sigma})}{\max_K L(\boldsymbol{\Theta}, \boldsymbol{\Sigma})}.$$

Then we can express

$$\begin{aligned} -2\log\lambda &= \min_H\{-2\log L(\boldsymbol{\Theta}, \boldsymbol{\Sigma})\} - \min_K\{-2\log L(\boldsymbol{\Theta}, \boldsymbol{\Sigma})\} \\ &= \min_H\{n\log|\boldsymbol{\Sigma}| + \mathrm{tr}(\mathbf{Y}-\mathbf{X}\boldsymbol{\Theta})'(\mathbf{Y}-\mathbf{X}\boldsymbol{\Theta})\} \\ &\quad -\min_K\{n\log|\boldsymbol{\Sigma}| + \mathrm{tr}(\mathbf{Y}-\mathbf{X}\boldsymbol{\Theta})'(\mathbf{Y}-\mathbf{X}\boldsymbol{\Theta})\}. \end{aligned}$$

Using Theorem 2.1, we can expressed as

$$\lambda^{2/n} \equiv \boldsymbol{\Lambda} = \frac{|n\hat{\boldsymbol{\Sigma}}_\Omega|}{|n\hat{\boldsymbol{\Sigma}}_\omega|}.$$

Here, $\hat{\boldsymbol{\Sigma}}_\Omega$ and $\hat{\boldsymbol{\Sigma}}_\omega$ are the maximum likelihood estimators of ΣΣ under the model (2.1) or K and H, respectively, which are given by

$$\begin{aligned} n\hat{\boldsymbol{\Sigma}}_\Omega &= (\mathbf{Y}-\mathbf{X}\hat{\boldsymbol{\Theta}}_\Omega)'(\mathbf{Y}-\mathbf{X}\hat{\boldsymbol{\Theta}}_\Omega), \quad \hat{\boldsymbol{\Theta}}_\Omega = (\mathbf{X}'\mathbf{X})^{-1}\mathbf{X}'\mathbf{Y} \\ &= \mathbf{Y}'(\mathbf{I}_n - \mathbf{P}_\Omega)\mathbf{Y} \end{aligned}$$

and

$$\begin{aligned} n\hat{\boldsymbol{\Sigma}}_\omega &= (\mathbf{Y}-\mathbf{X}_1\hat{\boldsymbol{\Theta}}_{1\omega})'(\mathbf{Y}-\mathbf{X}_1\hat{\boldsymbol{\Theta}}_{1\omega}), \quad \hat{\boldsymbol{\Theta}}_{1\omega} = (\mathbf{X}_1'\mathbf{X}_1)^{-1}\mathbf{X}_1'\mathbf{Y} \\ &= \mathbf{Y}'(\mathbf{I}_n - \mathbf{P}_\omega)\mathbf{Y} \end{aligned}$$

Summarizing these results, we have the following theorem.

Theorem 2.2 *Let* $\lambda = \Lambda^{n/2}$ *be the* LRC *for testing* H *in* (2.2). *Then,* Λ *is expressed as*

$$\Lambda = \frac{|\mathbf{S}_e|}{|\mathbf{S}_e + \mathbf{S}_h|},$$

where

$$\mathbf{S}_e = n\hat{\boldsymbol{\Sigma}}_\Omega, \quad \mathbf{S}_h = n\hat{\boldsymbol{\Sigma}}_\omega - n\hat{\boldsymbol{\Sigma}}_\Omega,$$

and $\mathbf{S}_\Omega$ *and* $\mathbf{S}_\omega$ *are given by* (2.4) *and* (2.5), *respectively.*

The matrices $\mathbf{S}_e$ and $\mathbf{S}_h$ in the testing problem are called the sums of squares and products (SSP) matrices due to the error and the hypothesis, respectively. We consider the distribution of Λ. If a $p \times p$ random matrix **W** is expressed as

$$\mathbf{W} = \sum_{j=1}^{n} z_j z_j'$$

where $z_j \sim N_p(\mu_j, \Sigma)$ and $z_1, \ldots, z_n$ are independent, **W** is said to have a noncentral Wishart distribution with n degrees of freedom, covariance matrix Σ,Σ, and noncentrality matrix $\Delta = \mu_1\mu_1' + \cdots + \mu_n\mu_n'$. We write that $\mathbf{W} \sim W_p(n, \Sigma; \Delta)$. In the special case Δ=**O, W** is said to have a Wishart distribution, denoted by $\mathbf{W} \sim W_p(n, \Sigma)$.

Theorem 2.3 (multivariate Cochran theorem) *Let* $\mathbf{Y}=(\mathbf{y}_1, \dots, \mathbf{y}_n)'$, *where* $\mathbf{y}_i \sim N_p(\mu_i, \Sigma)$, $i = 1, \dots, n$ *and* $\mathbf{y}_1, \dots, \mathbf{y}_n$ *are independent. Let* **A**, $\mathbf{A}_1$, *and* $\mathbf{A}_2$ *be* $n \times n$ *symmetric matrices. Then:*

1. $\mathbf{Y}'\mathbf{A}\mathbf{Y} \sim W_p(k, \Sigma; \Omega) \Leftrightarrow \mathbf{A}^2=\mathbf{A}, \quad \mathrm{tr}\mathbf{A}=k, \quad \Omega = E(\mathbf{Y})'\mathbf{A}E(\mathbf{Y}).$

2. $\mathbf{Y}'\mathbf{A}_1\mathbf{Y}$ *and* $\mathbf{Y}'\mathbf{A}^2\,\mathbf{Y}$ *are independent* $\Leftrightarrow \mathbf{A}_1\mathbf{A}_2 = \mathbf{O}$.

For a proof of multivariate Cochran theorem, see, e.g. [3, 6–8]. Let **B** and **W** be independent random matrices following the Wishart distribution $W_p(q, \Sigma)$ and $W_p(n, \Sigma)$, respectively, with $n \geq p$. Then, the distribution of

$$\Lambda = \frac{|\mathbf{W}|}{|\mathbf{B}+\mathbf{W}|}$$

is said to be the p-dimensional Lambda distribution with (q, n)-degrees of freedom and is denoted by $\Lambda_p(q, n)$. For distributional results of $\Lambda_p(q, n)$, see [1, 3].

By using multivariate Cochran's theorem, we have the following distributional results:

Theorem 2.4 *Let* $\mathbf{S}_e$ *and* $\mathbf{S}_h$ *be the random matrices in* (2.7). *Let* Λ *be the* Λ*-statistic defined by* (2.6). *Then,*

1. $\mathbf{S}_e$ *and* $\mathbf{S}_h$ *are independently distributed as a Wishart distribution* $W_p(n-k,\Sigma)$ *and a noncentral Wishart distribution* $W_p(k-j,\Sigma;\Delta)$, *respectively, where*

$$\Delta = (\mathbf{X}\Theta)'(\mathbf{P}_{\mathbf{X}} - \mathbf{P}_{\mathbf{X}_1})\mathbf{X}\Theta.$$

2. *Under H, the statistic* Λ *is distributed as a lambda distribution* $\Lambda_p(k-j, n-k)$.

Proof. Note that $\mathbf{P}_\Omega = \mathbf{P}_{\mathbf{X}} = \mathbf{X}(\mathbf{X}'\mathbf{X})^{-1}\mathbf{X}'$, $\mathbf{P}_\omega = \mathbf{P}_{X1} = \mathbf{X}_1(\mathbf{X}'_1\mathbf{X}_1)^{-1}\mathbf{X}_1'$ and $\mathbf{P}_\Omega\mathbf{P}_\omega = \mathbf{P}_\omega\mathbf{P}_\Omega$. By multivariate Cochran's theorem the first result (1) follows by checking that

$$(\mathbf{I}_n - \mathbf{P}_\Omega)^2 = (\mathbf{I}_n - \mathbf{P}_\Omega), \quad (\mathbf{P}_\Omega - \mathbf{P}_\omega)^2 = (\mathbf{P}_\Omega - \mathbf{P}_\omega),$$
$$(\mathbf{I}_n - \mathbf{P}_\Omega)(\mathbf{P}_\Omega - \mathbf{P}_\omega) = \mathbf{O}.$$

The second result (2) follows by showing that Δ_0 =O, where Δ_0 is the Δ under *H*. This is seen that

$$\Delta_0 = (\mathbf{X}_1\Theta_1)'(\mathbf{P}_\Omega - \mathbf{P}_\omega)(\mathbf{X}_1\Theta_1) = \mathbf{O},$$

since $\mathbf{P}_\Omega\mathbf{X}_1 = \mathbf{P}_\omega\mathbf{X}_1 = \mathbf{X}_1$.

The matrices $\mathbf{S}_e$ and $\mathbf{S}_h$ in (2.7) are defined in terms of $n \times n$ matrices $\mathbf{P}_\Omega$ and $\mathbf{P}_\omega$. It is important to give expressions useful for their numerical computations. We have the following expressions:

$$\mathbf{S}_e = \mathbf{Y}'\mathbf{Y} - \mathbf{Y}'\mathbf{X}(\mathbf{X}'\mathbf{X})^{-1}\mathbf{X}'\mathbf{Y}, \quad \mathbf{S}_h = \mathbf{Y}'\mathbf{X}(\mathbf{X}'\mathbf{X})^{-1}\mathbf{X}'\mathbf{Y} - \mathbf{Y}'\mathbf{X}_1(\mathbf{X}_1'\mathbf{X}_1)^{-1}\mathbf{X}_1'\mathbf{Y}.$$

Suppose that x_1 is 1 for all subjects, i.e., x_1 is an intercept term. Then, we can express these in terms of the SSP matrix of $(\mathbf{y}', \mathbf{x}')'$ defined by

$$\mathbf{S} = \sum_{i=1}^{n} \begin{pmatrix} y_i - \bar{y} \\ x_i - \bar{x} \end{pmatrix} \begin{pmatrix} y_i - \bar{y} \\ x_i - \bar{x} \end{pmatrix} = \begin{pmatrix} \mathbf{S}_{yy} & \mathbf{S}_{yx} \\ \mathbf{S}_{xy} & \mathbf{S}_{xx} \end{pmatrix},$$

where $\bar{\mathrm{y}}$ and $\bar{\mathrm{x}}$ are the sample mean vectors. Along the partition of $\mathrm{x}=(\mathbf{x}'_1,\mathbf{x}'_2)'$, we partition $\mathbf{S}$ as

$$\mathbf{S} = \begin{pmatrix} \mathbf{S}_{yy} & \mathbf{S}_{y1} & \mathbf{S}_{y2} \\ \mathbf{S}_{y1} & \mathbf{S}_{11} & \mathbf{S}_{12} \\ \mathbf{S}_{y2} & \mathbf{S}_{21} & \mathbf{S}_{22} \end{pmatrix}.$$

Then,

$$\mathbf{S}_e = \mathbf{S}_{yy\cdot x}, \quad \mathbf{S}_h = \mathbf{S}_{y2\cdot 1}\mathbf{S}_{22\cdot 1}^{-1}\mathbf{S}_{2y\cdot 1}.$$

Here, we use the notation $\mathbf{S}_{yy\cdot x} = \mathbf{S}_{yy} - \mathbf{S}_{yx}\mathbf{S}_{xx}^{-1}\mathbf{S}_{xy}$, $\mathbf{S}_{y2\cdot 1} = \mathbf{S}_{y2} - \mathbf{S}_{y1}\mathbf{S}_{11}^{-1}\mathbf{S}_{1y}$, etc. These are derived in the next section by using projection matrices.

3. IDEMPOTENT MATRICES AND MAX-MINI PROBLEMS

In the previous section, we have seen that idempotent matrices play an important role on statistical inference in multivariate regression model. In fact, letting $\mathrm{E}(\mathbf{Y}) = \eta = (\eta_1, \ldots, \eta_p)$ consider a model satisfying

$$\boldsymbol{\eta}_i \in \Omega = \mathcal{R}[\mathbf{R}], \quad \text{for all } i = 1, \ldots, p.$$

Then the MLE of Θ is $\hat{\Theta} = (X'X)^{-1}X'Y$, and hence the MLE of η is denoted by

$$\hat{\boldsymbol{\eta}}_\Omega = \mathbf{X}\hat{\boldsymbol{\Theta}} = \mathbf{P}_\Omega\mathbf{Y}.$$

Here, $\mathbf{P}_\Omega = \mathbf{X}(\mathbf{X}'\mathbf{X})^{-1}\mathbf{X}'$. Further, the residual sums of squares and products (RSSP) matrix is expressed as

$$\mathrm{S}_\Omega = (\mathbf{Y} - \hat{\boldsymbol{\eta}}_\Omega)'(\mathbf{Y} - \hat{\boldsymbol{\eta}}_\Omega) = \mathbf{Y}'(\mathbf{I}_n - \mathbf{P}_\Omega)\mathbf{Y}.$$

Under the hypothesis (2.2), the spaces η_i's belong are the same and are given by $\omega = \mathcal{R}[\mathbf{X}_1]$. Similarly, we have

$$\hat{\boldsymbol{\eta}}_\omega = \mathbf{X}\hat{\boldsymbol{\Theta}}_\omega = \mathbf{P}_\omega\mathbf{Y},$$
$$\mathbf{S}_\omega = (\mathbf{Y} - \hat{\boldsymbol{\eta}}_\omega)'(\mathbf{Y} - \hat{\boldsymbol{\eta}}_\omega) = \mathbf{Y}'(\mathbf{I}_n - \mathbf{P}_\omega)\mathbf{Y},$$

where $\hat{\Theta}_\omega = (\hat{\Theta}_{1\omega}', \mathrm{O})'$ and $\hat{\Theta}_{1\omega} = (X_1'X_1)^{-1}X_1'Y$. The LR criterion is based on the following decomposition of SSP matrices;

$$\begin{aligned} \mathbf{S}_\omega = \mathbf{Y}'(\mathbf{I}_n - \mathbf{P}_\omega)\mathbf{Y} &= \mathbf{Y}'(\mathbf{I}_n - \mathbf{P}_\Omega)\mathbf{Y} + \mathbf{Y}'(\mathbf{P}_\Omega - \mathbf{P}_\omega)\mathbf{Y} \\ &= \mathbf{S}_e + \mathbf{S}_h. \end{aligned}$$

The degrees of freedom in the Λ distribution $\Lambda_p(f_h, f_e)$ are given by

$$f_e = n - \dim[\Omega], \quad f_h = k - j = \dim[\Omega] - \dim[\omega].$$

In general, an $n \times n$ matrix $\mathbf{P}$ is called idempotent if $\mathbf{P}^2 = \mathbf{P}$. A symmetric and idempotent matrix is called projection matrix. Let $\mathbf{R}^n$ be the n dimensional Euclid space, and Ω be a subspace in $\mathbf{R}^n$. Then, any $n \times 1$ vector $\mathbf{y}$ can be uniquely decomposed into direct sum, i.e.,

$$y = u + v, \quad u \in \Omega, \quad v \in \Omega^{\perp},$$

where $\Omega^{\perp}$ is the orthocomplement space. Using decomposition (3.2), consider a mapping

$$\mathbf{P}_{\Omega} : y \to u, \quad \text{i.e. } \mathbf{P}_{\Omega} y = u.$$

The mapping is linear, and hence it is expressed as a matrix. In this case, $\mathbf{u}$ is called the orthogonal projection of $\mathbf{y}$ into Ω, and $\mathbf{P}_{\Omega}$ is also called the orthogonal projection matrix to Ω. Then, we have the following basic properties:

(P1) $\mathbf{P}_{\Omega}$ is uniquely defined;

(P2) $\mathbf{I}_n - \mathbf{P}_{\Omega}$ is the projection matrix to $\Omega^{\perp}$;

(P3) $\mathbf{P}_{\Omega}$ is a symmetric idempotent matrix;

(P4) $\mathcal{R}[\mathbf{P}_{\Omega}] = \Omega$, and $\dim[\Omega] = tr\mathbf{P}_{\Omega}$;

Let ω be a subset of Ω. Then, we have the following properties:

(P5) $\mathbf{P}_{\Omega}\mathbf{P}_{\omega} = \mathbf{P}_{\omega}\mathbf{P}_{\Omega} = \mathbf{P}_{\omega}$.

(P6) $\mathbf{P}_{\Omega} - \mathbf{P}_{\omega} = \mathbf{P}_{\omega^{\perp} \cap \Omega}$, where $\omega^{\perp}$ is the orthocomplement space of ω.

(P7) Let $\mathbf{B}$ be a $q \times n$ matrix, and let $N(\mathbf{B}) = \{\mathbf{y};\ \mathbf{B}\mathbf{y} = \mathbf{0}\}$. If $\omega = N[\mathbf{B}] \cap \Omega$, then $\omega^{\perp} \cap \Omega = R[\mathbf{P}_{\Omega}\mathbf{B}']$.

For more details, see, e.g. [3, 7, 9, 10].

The MLEs and LRC in multivariate regression model are derived by using the following theorem.

Theorem 3.1

1. *Consider a function of* $f(\Sigma) = \log|\Sigma| + \text{tr}\Sigma^{-1}\mathbf{S}$ *of* $p \times p$ *positive definite matrix. Then,* $f(\Sigma)$ *takes uniquely the minimum at* $\Sigma = \mathbf{S}$*, and the minimum value is given by*

$$\min_{\mathbf{\Sigma} > \mathbf{O}} f(\mathbf{\Sigma}) = f(\mathbf{S}) + p.$$

2. *Let* $\mathbf{Y}$ *be an* $n \times p$ *known matrix and* $\mathbf{X}$ *an* $n \times k$ *known matrix of rank k. Consider a function of* $p \times p$ *positive definite matrix* Σ *and* $k \times p$ *matrix* $\Theta = (\theta_{ij})$ *given by*

$$g(\mathbf{\Theta}, \mathbf{\Sigma}) = m \log|\Sigma| + \text{tr}\Sigma^{-1}(\mathbf{Y} - \mathbf{X\Theta})'(\mathbf{Y} - \mathbf{X\Theta}),$$

where $m > 0$, $-\infty < \theta_{ij} < \infty$, *for* $i = 1, \ldots, k$; $j = 1, \ldots, p$. *Then,* $g(\Theta, \Sigma)$ *takes the minimum at*

$$\mathbf{\Theta} = \hat{\mathbf{\Theta}} = (\mathbf{X}'\mathbf{X})^{-1}\mathbf{X}'\mathbf{Y}, \quad \mathbf{\Sigma} = \hat{\mathbf{\Sigma}} = \frac{1}{m}\mathbf{Y}'(\mathbf{I}_n - \mathbf{P}_{\mathbf{X}})\mathbf{Y},$$

and the minimum value is given by $m \log|\hat{\mathbf{\Sigma}}| + mp$.

Proof. Let $\ell_1, ..., \ell_p$ be the characteristic roots of $\Sigma^{-1}\mathbf{S}$. Note that the characteristic roots of $\Sigma^{-1}\mathbf{S}$ and $\Sigma^{-1/2}\mathbf{S}\Sigma^{-1/2}$ are the same. The latter matrix is positive definite, and hence we may assume $\ell_1 \geq \cdots \geq \ell_p > 0$. Then

$$\begin{aligned} f(\mathbf{\Sigma}) - f(\mathbf{S}) &= \log|\mathbf{\Sigma}\mathbf{S}^{-1}| + tr(\mathbf{\Sigma}^{-1}\mathbf{S}) - p \\ &= -\log|\mathbf{\Sigma}^{-1}\mathbf{S}| + tr(\mathbf{\Sigma}^{-1}\mathbf{S}) - p \\ &= \sum_{i=1}^{p}(-\log \ell_i + \ell_i - 1) \geq 0. \end{aligned}$$

The last inequality follows from $x - 1 \geq \log x$ $(x > 0)$. The equality holds if and only if $\ell_1 = \cdots = \ell_p = 1 \Leftrightarrow \Sigma = \mathbf{S}$.

Next, we prove 2. we have

$$\begin{aligned} &\mathrm{tr}\mathbf{\Sigma}^{-1}(\mathbf{Y} - \mathbf{X}\mathbf{\Theta})'(\mathbf{Y} - \mathbf{X}\Theta) \\ &= \mathrm{tr}\mathbf{\Sigma}^{-1}(\mathbf{Y} - \mathbf{X}\hat{\mathbf{\Theta}})'(\mathbf{Y} - \mathbf{X}\hat{\mathbf{\Theta}}) + \mathrm{tr}\mathbf{\Sigma}^{-1}\{\mathbf{X}(\hat{\mathbf{\Theta}} - \mathbf{\Theta})\}'\mathbf{X}(\hat{\mathbf{\Theta}} - \mathbf{\Theta}) \\ &\geq \mathrm{tr}\mathbf{\Sigma}^{-1}\mathbf{Y}'(\mathbf{I}_n - \mathbf{P}_{\mathbf{X}})\mathbf{Y}. \end{aligned}$$

The first equality follows from that $\mathrm{Y} - \mathrm{X}\Theta = \mathrm{Y} - \mathrm{X}\hat{\Theta} + \mathrm{X}(\hat{\Theta} - \Theta)$ and $(\mathrm{Y} - \mathrm{X}\hat{\Theta})'\mathrm{X}(\hat{\Theta} - \Theta) = \mathrm{O}$. In the last step, the equality holds when $\Theta = \hat{\Theta}$. The required result is obtained by noting that Θ^Θ^ does not depend on ΣΣ and combining this result with the first result 1.

Theorem 3.2 *Let* $\mathbf{X}$ *be an* $n \times k$ *matrix of rank k, and let* $\Omega = \mathcal{R}[\mathbf{X}]$ *which is defined also by the set* $\{\mathbf{y} : \mathbf{y} = \mathbf{X}\theta\}$*, where* θ *is a* $k \times 1$ *unknown parameter vector. Let* $\mathbf{C}$ *be a* $c \times k$ *matrix of rank c, and define* ω *by the set* $\{\mathbf{y} : \mathbf{y} = \mathrm{X}\theta, \ \mathbf{C}\theta = \mathbf{0}\}$*. Then,*

1. $\mathbf{P}_\Omega = \mathbf{X}(\mathbf{X}'\mathbf{X})^{-1}\mathbf{X}'$.
2. $\mathbf{P}_\Omega - \mathbf{P}_\omega = \mathbf{X}(\mathbf{X}'\mathbf{X})^{-1}\mathbf{C}'\{\mathbf{C}(\mathbf{X}'\mathbf{X})^{-1}\mathbf{C}\}^{-1}\mathbf{C}(\mathbf{X}'\mathbf{X})^{-1}\mathbf{X}'$.

Proof. 1 Let $\hat{\mathbf{y}} = \mathbf{X}(\mathbf{X}'\mathbf{X})^{-1}\mathbf{X}'$ and consider a decomposition $\mathbf{y} = \hat{\mathbf{y}} + (\mathbf{y} - \hat{\mathbf{y}})$. Then, $\hat{\mathbf{y}}'(\mathbf{y} - \hat{\mathbf{y}}) = \mathbf{0}$. Therefore, $\mathbf{P}_\Omega\mathbf{y} = \hat{\mathbf{y}}$ and hence $\mathbf{P}_\Omega = \mathbf{X}(\mathbf{X}'\mathbf{X})^{-1}\mathbf{X}'$.

2. Since $\mathbf{C}\theta = \mathbf{C}(\mathbf{X}'\mathbf{X})^{-1}\mathbf{X}' \cdot \mathbf{X}\theta$, we can write $\omega = \mathrm{N}[\mathbf{B}] \cap \Omega$, where $\mathbf{B} = \mathbf{C}(\mathbf{X}'\mathbf{X})^{-1}\mathbf{X}'$. Using (P7),

$$\omega^{\perp} \cap \Omega = \mathcal{R}[\mathbf{P}_\Omega\mathbf{B}'] = \mathcal{R}[\mathbf{X}(\mathbf{X}'\mathbf{X})^{-1}\mathbf{C}'].$$

The final result is obtained by using 1 and (P7).

Consider a special case $\mathbf{C} = (\mathbf{O} \ \ \mathbf{I}_{k-q})$. Then $\omega = \mathcal{R}[\mathbf{X}_1]$, where $\mathbf{X} = (\mathbf{X}_1 \ \ \mathbf{X}_2)$, $\mathbf{X}_1 : n \times q$. We have the following results:

$$\begin{aligned} &\omega^{\perp} \cap \Omega = \mathcal{R}[(\mathbf{I}_n - \mathbf{P}_{\mathbf{X}_1})\mathbf{X}_2], \\ &\mathbf{P}_{\omega^{\perp} \cap \Omega} = (\mathbf{I}_n - \mathbf{P}_{\mathbf{X}_1})\mathbf{X}_2\{\mathbf{X}_2'(\mathbf{I}_n - \mathbf{P}_{\mathbf{X}_1})\mathbf{X}_2\}^{-1}\mathbf{X}_2'(\mathbf{I}_n - \mathbf{P}_{\mathbf{X}_1}). \end{aligned}$$

The expressions (2.11) for $\mathbf{S}_e$ and $\mathbf{S}_h$ in terms of $\mathbf{S}$ can be obtained from projection matrices based on

$$\Omega = \mathcal{R}[\mathbf{X}] = \mathcal{R}[\mathbf{1}_n] + \mathcal{R}[(\mathbf{I}_n - \mathbf{P}_{\mathbf{1}_n})\mathbf{X}],$$

$$\omega^{\perp} \cap \Omega = \mathcal{R}[(\mathbf{I}_n - \mathbf{P}_{\mathbf{1}_n} - \mathbf{P}_{(\mathbf{I}_n - \mathbf{P}_{\mathbf{1}_n})\mathbf{X}_1})\mathbf{X}_2].$$

4. GENERAL LINEAR HYPOTHESIS

In this section, we will examine the process of testing a broad linear hypothesis.

$$H_g : \mathbf{C}\Theta\mathbf{D} = \mathbf{O},$$

against alternatives K_g : $\mathbf{C}\Theta\mathbf{D} \neq \mathbf{O}$ under a multivariate linear model given by (2.1), where $\mathbf{C}$ is a $c \times k$ given matrix with rank c and $\mathbf{D}$ is a $p \times d$ given matrix with rank d. When $\mathbf{C} = (\mathbf{O}\ \ \mathbf{I}_{k-j})$ and $\mathbf{D} = \mathbf{I}_p$, the hypothesis H_g becomes $H : \Theta_2 = \mathbf{O}$.

For the derivation of LR test of (4.1), we can use the following conventional approach: If $\mathbf{U}=\mathbf{YD}$, then the rows of $\mathbf{U}$ are independent and normally distributed with the identical covariance matrix $\mathbf{D}'\Sigma\mathbf{D}$, and

$$\mathrm{E}(\mathbf{U}) = \mathbf{X}\Xi,$$

where $\Xi = \Theta\mathbf{D}$. The hypothesis (4.1) is expressed as

$$H_g : \mathbf{C}\Xi = \mathbf{O}.$$

Applying a general theory for testing H_g in (2.1), we have the LRC λ:

$$\lambda^{2/n} = \Lambda = \frac{|\mathbf{S}_e|}{|\mathbf{S}_e + \mathbf{S}_h|},$$

where

$$\begin{aligned}\mathbf{S}_e &= \mathbf{U}'(\mathbf{I}_n - \mathbf{P}_{\mathbf{X}})\mathbf{U}\\ &= \mathbf{D}'\mathbf{Y}'(\mathbf{I}_n - \mathbf{P}_{\mathbf{A}})\mathbf{YD},\end{aligned}$$

and

$$\begin{aligned}\mathbf{S}_h &= (\mathbf{C}(\mathbf{X}'\mathbf{X})^{-1}\mathbf{X}'\mathbf{U})'\{\mathbf{C}(\mathbf{X}'\mathbf{X})^{-1}\mathbf{C}'\}^{-1}\mathbf{C}(\mathbf{X}'\mathbf{X})^{-1}\mathbf{X}'\mathbf{U},\\ &= (\mathbf{C}(\mathbf{X}'\mathbf{X})^{-1}\mathbf{X}'\mathbf{YD})'\{\mathbf{C}(\mathbf{X}'\mathbf{X})^{-1}\mathbf{C}'\}^{-1}\mathbf{C}(\mathbf{X}'\mathbf{X})^{-1}\mathbf{X}'\mathbf{Y}D.\end{aligned}$$

Theorem 4.1 *The statistic Λ in (4.4) is an LR statistic for testing (4.1) under (2.1). Further, under H_g, $\Lambda \sim \Lambda_d(c, n - k)$.*

Proof. Let $\mathbf{G} = (\mathbf{G}_1\ \ \mathbf{G}_2)$ be a $p \times p$ matrix such that $\mathbf{G}_1 = \mathbf{D}$, $\mathbf{G}'_1\mathbf{G}_2=\mathbf{O}$, and $|\mathbf{G}| \neq 0$. Consider a transformation from $\mathbf{Y}$ to $(\mathbf{U}\ \mathbf{V}) = \mathbf{Y}(\mathbf{G}_1\ \mathbf{G}_2)$.

Then the rows of $(\mathbf{U}\ \mathbf{V})$ are independently normal with the same covariance matrix

$$\Psi = \mathbf{G}'\Sigma\mathbf{G} = \begin{pmatrix}\Psi_{11} & \Psi_{12}\\ \Psi_{21} & \Psi_{22}\end{pmatrix}, \quad \Psi_{12} : d \times (p - d),$$

and

$$E[(\mathrm{U}\ \mathrm{V})] = \mathbf{X}\Theta(\mathbf{G}_1\ \mathbf{G}_2)$$
$$= \mathbf{X}(\Xi\ \Delta), \quad \Xi = \Theta\mathbf{G}_1, \Delta = \Theta\mathbf{G}_2.$$

The conditional of **V** given **U** is normal. The rows of **V** given **U** are independently normal with the same covariance matrix Ψ11·2Ψ11·2 , and

$$\mathrm{E}(\mathrm{V} \mid \mathrm{U}) = \mathbf{X\Delta} + (\mathrm{U} - \mathbf{X\Xi})\boldsymbol{\Gamma}$$
$$= \mathbf{X\Delta}^* + \mathrm{U}\Gamma,$$

where $\Delta^* = \Delta - \Xi\Gamma$ and $\Gamma = \Psi_{11}^{-1}\Psi_{12}$. We see that the maximum likelihood of **V** given **U** does not depend on the hypothesis. Therefore, an LR statistic is obtained from the marginal distribution of U,U, which implies the results required.

5. ADDITIONAL INFORMATION TESTS FOR RESPONSE VARIABLES

We consider a multivariate regression model with an intercept term x_0 and k explanatory variables $x_1, \ldots, x_k$ as follows.

$$\mathrm{Y} = \mathbf{1}\boldsymbol{\theta}' + \mathrm{X}\Theta + \mathbf{E},$$

where **Y** and **X** are the observation matrices on $\mathbf{y} = (y_1, \ldots, y_p)'$ and $\mathbf{x} = (x_1, \ldots, x_k)'$. We assume that the error matrix **E** has the same property as in (2.1), and rank $(\mathbf{1}_n\ \mathbf{X}) = k + 1$. Our interest is to test a hypothesis $H_{2\cdot1}$ on no additional information of $\mathbf{y}_2 = (y_{q+1}, \ldots, y_p)'$ in presence of $\mathbf{y}_1 = (y_1, \ldots, y_q)'$.

Along the partition of **y** into $(\mathbf{y}_1', \mathbf{y}_2')$ let Y,Y, θ, Θ,Θ, and ΣΣ partition as

$$\mathrm{Y} = (\mathrm{Y}_1\ \mathrm{Y}_2), \quad \boldsymbol{\Theta} = (\boldsymbol{\Theta}_1\ \boldsymbol{\Theta}_2),$$
$$\boldsymbol{\theta} = \begin{pmatrix} \boldsymbol{\theta}_1 \\ \boldsymbol{\theta}_2 \end{pmatrix}, \quad \boldsymbol{\Sigma} = \begin{pmatrix} \boldsymbol{\Sigma}_{11} & \boldsymbol{\Sigma}_{12} \\ \boldsymbol{\Sigma}_{21} & \boldsymbol{\Sigma}_{22} \end{pmatrix}.$$

The conditional distribution of $\mathbf{Y}_2$ given $\mathbf{Y}_1$ is normal with mean

$$\mathrm{E}(\mathrm{Y}_2 \mid \mathrm{Y}_1) = \mathbf{1}\boldsymbol{\theta}_2' + \mathrm{X}\Theta_2 + (\mathrm{Y}_1 - \mathbf{1}_n\boldsymbol{\theta}_1' - \mathbf{X}\boldsymbol{\Theta}_1)\boldsymbol{\Sigma}_{11}^{-1}\boldsymbol{\Sigma}_{12}$$
$$= \mathbf{1}_n\tilde{\boldsymbol{\theta}}_{02}' + \mathbf{X}\tilde{\boldsymbol{\Theta}}_2 + \mathrm{Y}_1\boldsymbol{\Sigma}_{11}^{-1}\boldsymbol{\Sigma}_{12},$$

and the conditional covariance matrix is expressed as

$$\mathrm{Var}[\mathrm{vec}\,(\mathrm{Y}_2 \mid \mathrm{Y}_1)] = \boldsymbol{\Sigma}_{22\cdot1} \otimes \mathbf{I}_n,$$

where $\Sigma_{22\cdot1} = \Sigma_{22} - \Sigma_{21}\Sigma_{11}^{-1}\Sigma_{12}$, and

$$\tilde{\boldsymbol{\theta}}_2' = \boldsymbol{\theta}_2' - \boldsymbol{\theta}_1'\boldsymbol{\Sigma}_{11}^{-1}\boldsymbol{\Sigma}_{12}, \quad \tilde{\boldsymbol{\Theta}}_2 = \boldsymbol{\Theta}_2 - \boldsymbol{\Theta}_1\boldsymbol{\Sigma}_{11}^{-1}\boldsymbol{\Sigma}_{12}.$$

Here, for an $n \times p$ matrix $\mathbf{Y} = (\mathbf{y}_{(1)}, \ldots, \mathbf{y}_{(p,)})$ *vec* $(\mathbf{Y})$ means an np-vector $(\mathbf{y}'_{(1)}, \ldots, \mathbf{y}'_{(p)})'$. Now we define the hypothesis $H_{2\cdot1}$ as

$$H_{2\cdot 1}: \mathbf{\Theta}_2 = \mathbf{\Theta}_1 \mathbf{\Sigma}_{11}^{-1} \mathbf{\Sigma}_{12} \Leftrightarrow \tilde{\mathbf{\Theta}}_2 = \mathrm{O}.$$

The hypothesis $H_{2\cdot 1}$ means that $\mathbf{y}_2$ after removing the effects of $\mathbf{y}_1$ does not depend on $\mathbf{x}$. In other words, the relationship between $\mathbf{y}_2$ and $\mathbf{x}$ can be described by the relationship between $\mathbf{y}_1$ and $\mathbf{x}$. In this sense, $\mathbf{y}_2$ is redundant in the relationship between $\mathbf{y}$ and $\mathbf{x}$.

The LR criterion for testing the hypothesis $H_{2\cdot 1}$ against alternatives $K_{2\cdot 1}: \tilde{\Theta}_{2\cdot 1} \neq \mathrm{O}$ can be obtained through the following steps.

(D1) The density function of $\mathbf{Y}=(\mathbf{Y}_1\mathbf{Y}_2)$ can be expressed as the product of the marginal density function of $\mathbf{Y}_1$ and the conditional density function of $\mathbf{Y}_2$ given $\mathbf{Y}_1$. Note that the density functions of $\mathbf{Y}_1$ under $H_{2\cdot 1}$ and $K_{2\cdot 1}$ are the same.

(D2) The spaces spanned by each column of $\mathrm{E}(\mathbf{Y}_2|\mathbf{Y}_1)$ are the same, and let the spaces under $K_{2\cdot 1}$ and $H_{2\cdot 1}$ denote by Ω and ω, respectively. Then

$$\Omega = \mathcal{R}[(\mathbf{1}_n\ \mathrm{Y}_1\ \mathbf{X})], \quad \omega = \mathcal{R}[(\mathbf{1}_n\ \mathrm{Y}_1)].$$

and $\dim(\Omega) = q + k + 1$, $\dim(\omega) = k + 1$.

(D3) The likelihood ratio criterion λ is expressed as

$$\lambda^{2/n} = \Lambda = \frac{|\mathbf{S}_\Omega|}{|\mathbf{S}_\omega|} = \frac{|\mathbf{S}_\Omega|}{|\mathbf{S}_\Omega + (\mathbf{S}_\omega - \mathbf{S}_\Omega)|}.$$

where $\mathbf{S}_\Omega = \mathrm{Y}_2'(\mathrm{I}_n - \mathrm{P}_\Omega)\mathrm{Y}_2$ and $\mathbf{S}_\omega = \mathrm{Y}_2'(\mathrm{I}_n - \mathrm{P}_\omega)\mathrm{Y}_2$.

(D4) Note that $\mathrm{E}(\mathrm{Y}_2 | \mathrm{Y}_1)'(\mathrm{P}_\omega - \mathrm{P}_\omega)\mathrm{E}(\mathrm{Y}_2 | \mathrm{Y}_1) = \mathbf{O}$ under $H_{2\cdot 1}$. The conditional distribution of Λ under $H_{2\cdot 1}$ is $\Lambda_{p-q}(k, n-q-k-1)$, and hence the distribution of Λ under $H_{2\cdot 1}$ is $\Lambda_{p-q}(k, n-q-k-1)$.

Note that the Λ statistic is defined through $\mathrm{Y}_2'(\mathrm{I}_n - \mathrm{P}_\Omega)\mathrm{Y}_2$ and $\mathrm{Y}_2'(\mathrm{P}_\Omega - \mathrm{P}_\omega)\mathrm{Y}_2$, which involve $n \times n$ matrices. We try to write these statistics in terms of the SSP matrix of $(\mathbf{y}', \mathbf{x}')'$ defined by

$$\mathbf{S} = \sum_{i=1}^{n} \begin{pmatrix} y_i - \bar{y} \\ x_i - \bar{x} \end{pmatrix} \begin{pmatrix} y_i - \bar{y} \\ x_i - \bar{x} \end{pmatrix}' = \begin{pmatrix} \mathbf{S}_{yy} & \mathbf{S}_{yx} \\ \mathbf{S}_{xy} & \mathbf{S}_{xx} \end{pmatrix},$$

where $\bar{\mathbf{y}}$ and $\bar{\mathbf{x}}$ are the sample mean vectors. Along the partition of $\mathbf{y}=(\mathbf{y}'_1, \mathbf{y}'_2)'$, we partition $\mathbf{S}$ as

$$\mathbf{S} = \begin{pmatrix} \mathbf{S}_{11} & \mathbf{S}_{12} & \mathbf{S}_{1x} \\ \mathbf{S}_{21} & \mathbf{S}_{22} & \mathbf{S}_{2x} \\ \mathbf{S}_{x1} & \mathbf{S}_{x2} & \mathbf{S}_{xx} \end{pmatrix}.$$

We can show that

$$\mathbf{S}_{\omega} = \mathbf{S}_{22\cdot1} = \mathbf{S}_{22} - \mathbf{S}_{21}\mathbf{S}_{11}^{-1}\mathbf{S}_{12},$$
$$\mathbf{S}_{\Omega} = \mathbf{S}_{22\cdot1x} = \mathbf{S}_{22\cdot x} - \mathbf{S}_{21\cdot x}\mathbf{S}_{11\cdot x}^{-1}\mathbf{S}_{12\cdot x}.$$

The first result is obtained by using

$$\omega = \mathcal{R}[\mathbf{1}_n] + \mathcal{R}[(\mathbf{I}_n - \mathbf{P}_{\mathbf{1}_n})\mathbf{Y}_1].$$

The second result is obtained by using

$$\begin{aligned}\Omega &= \mathcal{R}[\mathbf{1}_n] + \mathcal{R}[(\tilde{\mathbf{Y}}_1, \tilde{\mathbf{X}})] \\ &= \mathcal{R}[\mathbf{1}_n] + \mathcal{R}[(\mathbf{I}_n - \mathbf{P}_0)\mathbf{X}] + \mathcal{R}[(\mathbf{I}_n - \mathbf{P}_{\tilde{X}})(\mathbf{I}_n - \mathbf{P}_0)\mathbf{Y}_1],\end{aligned}$$

where $\tilde{\mathbf{Y}}_1 = (\mathbf{I}_n - \mathbf{P}_{\mathbf{1}_n})\mathbf{Y}_1$ and $\tilde{\mathbf{X}} = (\mathbf{I}_n - \mathbf{P}_{\mathbf{1}_n})\mathbf{X}$.

Summarizing the above results, we have the following theorem.

Theorem 5.1 *In the multivariate regression model* (5.1), *consider to test the hypothesis* $H_{2\cdot1}$ *in* (5.4) *against* $K_{2\cdot1}$. *Then the LR criterion* λ *is given by*

$$\lambda^{2/n} = \Lambda = \frac{|\mathbf{S}_{22\cdot1x}|}{|\mathbf{S}_{22\cdot1}|},$$

whose null distribution is $\Lambda_{p-q}(k, n - q - k - 1)$.

Note that $S_{22\cdot1}$ can be decomposed as

$$\mathbf{S}_{22\cdot1} = \mathbf{S}_{22\cdot1x} + \mathbf{S}_{2x\cdot1}\mathbf{S}_{xx\cdot1}^{-1}\mathbf{S}_{x2\cdot1}.$$

This decomposition is obtained by expressing $\mathbf{S}_{22\cdot1x}$ in terms of $\mathbf{S}_{22\cdot1}$, $\mathbf{S}_{2x\cdot1}$, $\mathbf{S}_{xx\cdot1}$, and $\mathbf{S}_{x2\cdot1}$ by using an inverse formula

$$\begin{pmatrix} \mathbf{H}_{11} & \mathbf{H}_{12} \\ \mathbf{H}_{21} & \mathbf{H}_{22} \end{pmatrix}^{-1} = \begin{pmatrix} \mathbf{H}_{11}^{-1} & \mathbf{O} \\ \mathbf{O} & \mathbf{O} \end{pmatrix} + \begin{pmatrix} -\mathbf{H}_{11}^{-1}\mathbf{H}_{12} \\ \mathbf{I} \end{pmatrix} \mathbf{H}_{22\cdot1}^{-1} \begin{pmatrix} -\mathbf{H}_{21}\mathbf{H}_{11}^{-1} & \mathbf{I} \end{pmatrix}.$$

The decomposition is expressed as

$$\mathbf{S}_{22\cdot1} - \mathbf{S}_{22\cdot1x} = \mathbf{S}_{2x\cdot1}\mathbf{S}_{xx\cdot1}^{-1}\mathbf{S}_{x2\cdot1}.$$

The result may be also obtained by the following algebraic method. We have

$$\begin{aligned}\mathbf{S}_{22\cdot1} - \mathbf{S}_{22\cdot1x} &= \mathbf{Y}_2'(\mathbf{P}_{\Omega} - \mathbf{P}_{\omega})\mathbf{Y}_2 \\ &= \mathbf{Y}_2'(\mathbf{P}_{\omega^{\perp}\cap\Omega})\mathbf{Y}_2.\end{aligned}$$

and

$$\Omega = \mathcal{R}[\mathbf{1}_n] + \mathcal{R}[(\tilde{\mathbf{Y}}_1\ \tilde{\mathbf{X}})], \quad \omega = \mathcal{R}[\mathbf{1}_n] + \mathcal{R}[\tilde{\mathbf{Y}}_1].$$

Therefore,

$$\begin{aligned}\omega^{\perp} \cap \Omega &= \mathcal{R}[(\mathbf{I}_n - \mathbf{P}_{\mathbf{1}_n} - \mathbf{P}_{\tilde{\mathbf{Y}}_1})(\tilde{\mathbf{Y}}_1\ \tilde{\mathbf{X}})] \\ &= \mathcal{R}[(\mathbf{I}_n - \mathbf{P}_{\mathbf{1}_n} - \mathbf{P}_{\tilde{\mathbf{Y}}_1})\tilde{\mathbf{X}}],\end{aligned}$$

which gives an expression for $\mathbf{P}_{\omega\perp\cap\Omega}$ by using Theorem 3.1 (1). This leads to (5.6).

6. TESTS IN DISCRIMINANT ANALYSIS

We consider q p-variate normal populations with common covariance matrix ΣΣ and the ith population having mean vector θ_i. Suppose that a sample of size n_i is available from the ith population, and let $\mathbf{y}_{ij}$ be the jth observation from the ith population. The observation matrix for all the observations is expressed as

$$\mathbf{Y} = (y_{11}, \ldots, y_{1n_1}, y_{21}, \ldots, y_{q1}, \ldots, y_{qn_q})'.$$

It is assumed that $\mathbf{y}_{ij}$ are independent, and

$$y_{ij} \sim \mathrm{N}(\boldsymbol{\theta}_i, \boldsymbol{\Sigma}),\ j = 1, \ldots, n_i;\ i = 1, \ldots, q.$$

The model is expressed as

$$\mathbf{Y} = \mathbf{A}\boldsymbol{\Theta} + \mathbf{E},$$

where

$$\mathbf{A} = \begin{pmatrix} \mathbf{1}_{n_1} & 0 & \cdots & 0 \\ 0 & \mathbf{1}_{n_2} & \cdots & 0 \\ \vdots & \vdots & \ddots & \vdots \\ 0 & 0 & \cdots & \mathbf{1}_{n_q} \end{pmatrix}, \quad \boldsymbol{\Theta} = \begin{pmatrix} \boldsymbol{\theta}_1' \\ \boldsymbol{\theta}_2' \\ \vdots \\ \boldsymbol{\theta}_q' \end{pmatrix}.$$

Here, the error matrix $\mathbf{E}$ has the same property as in (2.1).

First, we consider to test

$$H: \boldsymbol{\theta}_1 = \cdots = \boldsymbol{\theta}_q (= \boldsymbol{\theta}),$$

against alternatives $K: \theta_i \neq \theta_j$ for some i, j. The hypothesis can be expressed as

$$H: \mathbf{C}\boldsymbol{\Theta} = \mathbf{O}, \quad \mathbf{C} = (\mathbf{I}_{q-1}, -\mathbf{1}_{q-1}).$$

The tests including LRC are based on three basic statistics, the within-group SSP matrix W,W, the between-group SSP matrix B,B, and the total SSP matrix T given by

$$\mathsf{W} = \sum_{i=1}^{q}(n_i - 1)\mathsf{S}_i, \quad \mathsf{B} = \sum_{i=1}^{q} n_i(\bar{y}_i - \bar{y})(\bar{y}_i - \bar{y})'.$$

$$\mathsf{T} = \mathsf{B} + \mathsf{W} = \sum_{i=1}^{q}\sum_{j=1}^{n_i}(y_{ij} - \bar{y})(y_{ij} - \bar{y})'.$$

where $\bar{\mathsf{y}}_i$ and S_i are the mean vector and sample covariance matrix of the *i*th population, and $\bar{\mathsf{y}}$ is the total mean vector defined by $(1/n)\sum_{i=1}^{q} n_i\bar{\mathsf{y}}_i$ and $n = \sum_{i=1}^{q} n_i$. In general, **W** and **B** are independently distributed as a Wishart distribution $W_p(n-q,\Sigma)$ and a noncentral Wishart distribution $W_p(q-1, \Sigma; \Delta)$, respectively, where

$$\boldsymbol{\Delta} = \sum_{i=1}^{q} n_i(\boldsymbol{\theta}_i - \bar{\boldsymbol{\theta}})(\boldsymbol{\theta}_i - \bar{\boldsymbol{\theta}})',$$

where $\bar{\theta} = (1/n)\sum_{i=1}^{q-1} n_i\theta_i$. Then, the following theorem is well known.

Theorem 6.1 *Let* $\lambda = \Lambda^{n/2}$ *be the* LRC *for testing H in* (6.4). *Then,* Λ *is expressed as*

$$\Lambda = \frac{|\mathsf{W}|}{|\mathsf{W} + \mathsf{B}|} = \frac{|\mathsf{W}|}{|\mathsf{T}|},$$

where **W**, **B** , and **T** are given in (6.6). Further, under *H*, the statistic Λ is distributed as a lambda distribution $\Lambda_p(q - 1, n - q)$.

Now we shall show Theorem 6.1 by an algebraic method. It is easy to see that

$$\Omega = \mathcal{R}[\mathbf{A}], \quad \omega = \mathrm{N}[\mathbf{C}(\mathbf{A}'\mathbf{A})^{-1}\mathbf{A}'] \cap \Omega = \mathcal{R}[\mathbf{1}_n].$$

The last equality is also checked from that under *H*

$$\mathrm{E}[\mathsf{Y}] = \mathbf{A}\mathbf{1}_q\boldsymbol{\theta}' = \mathbf{1}_n\boldsymbol{\theta}'.$$

We have

$$\begin{aligned}\mathsf{T} &= \mathsf{Y}'(\mathbf{I}_n - \mathbf{P}_{\mathbf{1}_n})\mathsf{Y} \\ &= \mathsf{Y}'(\mathbf{I}_n - \mathbf{P}_{\mathbf{A}})\mathsf{Y} + \mathsf{Y}'(\mathbf{P}_{\mathbf{A}} - \mathbf{P}_{\mathbf{1}_n})\mathsf{Y} \\ &= \mathsf{W} + \mathsf{B}.\end{aligned}$$

Further, it is easily checked that

1. $(\mathbf{I}_n - \mathbf{P}_\mathbf{A})^2 = \mathbf{I}_n - \mathbf{P}_\mathbf{A}, \quad (\mathbf{P}_\mathbf{A} - \mathbf{P}_{\mathbf{1}_n})^2 = \mathbf{P}_\mathbf{A} - \mathbf{P}_{\mathbf{1}_n}.$

2. $(\mathbf{P}_\mathbf{A} - \mathbf{P}_{\mathbf{1}_n})(\mathbf{P}_\mathbf{A} - \mathbf{P}_{\mathbf{1}_n}) = \mathbf{O}.$

3. $f_e = \dim[\mathcal{R}[\mathbf{A}]^\perp] = \mathrm{tr}(\mathbf{I}_n - \mathbf{P}_\mathbf{A}) = n - q,$

$$f_h = \dim[\mathcal{R}[\mathbf{1}_n]^\perp \cap \mathcal{R}[\mathbf{A}]] = \mathrm{tr}(\mathbf{P}_\mathbf{A} - \mathbf{P}_{\mathbf{1}_n}) = q - 1.$$

Related to the test of H, we are interested in whether a subset of variables y_1, ..., y_p is sufficient for discriminant analysis, or the set of remainder variables has no additional information or is redundant. Without loss of generality, we consider the sufficiency of a subvector $\mathbf{y}_1 = (y_1, ..., y_k)'$ of $\mathbf{y}$, or redundancy of the remainder vector $\mathbf{y}_2 = (y_{k+1}, ..., y_p)'$. Consider to test

$$H_{2\cdot 1} : \boldsymbol{\theta}_{1;2\cdot 1} = \cdots = \boldsymbol{\theta}_{q;2\cdot 1} (= \boldsymbol{\theta}_{2\cdot 1}),$$

where

$$\boldsymbol{\theta}_i = \begin{pmatrix} \boldsymbol{\theta}_{i;1} \\ \boldsymbol{\theta}_{i;2} \end{pmatrix}, \quad \boldsymbol{\theta}_{i;1};\ k \times 1, \quad i = 1, \ldots, q,$$

and

$$\boldsymbol{\theta}_{i;2\cdot 1} = \boldsymbol{\theta}_{i;2} - \boldsymbol{\Sigma}_{21} \boldsymbol{\Sigma}_{11}^{-1} \boldsymbol{\theta}_{i;1}, \quad i = 1, \ldots, q.$$

The testing problem was considered by [11]. The hypothesis can be formulated in terms of Maharanobis distance and discriminant functions. For its details, see [12, 13]. To obtain a likelihood ratio for $H_{2\cdot 1}$, we partition the observation matrix as

$$\mathbf{Y} = (\mathbf{Y}_1 \quad \mathbf{Y}_2), \quad \mathbf{Y}_1 : n \times k.$$

Then the conditional distribution of $\mathbf{Y}_2$ given $\mathbf{Y}_1$ is normal such that the rows of $\mathbf{Y}_2$ are independently distributed with covariance matrix $\boldsymbol{\Sigma}_{22\cdot 1} = \boldsymbol{\Sigma}_{22} - \boldsymbol{\Sigma}_{21} \boldsymbol{\Sigma}_{11}^{-1} \boldsymbol{\Sigma}_{12}$, and the conditional mean is given by

$$\mathrm{E}(\mathbf{Y}_2 \mid \mathbf{Y}_1) = \mathbf{A} \boldsymbol{\Theta}_{2\cdot 1} + \mathbf{Y}_1 \boldsymbol{\Sigma}_{11}^{-1} \boldsymbol{\Sigma}_{12},$$

where $\Theta_{2\cdot 1} = (\theta_{1;2\cdot 1}, \ldots, \theta_{q;2\cdot 1})'$. The LRC for $H_{2\cdot 1}$ can be obtained by use of the conditional distribution, and following the steps (D1)–(D4) in Section 5. In fact, the spaces spanned by each column of $\mathrm{E}(\mathbf{Y}_2|\mathbf{Y}_1)$ are the same, and let the spaces under

$K_{2\cdot 1}$ and $H_{2\cdot 1}$ denote by Ω and ω, respectively. Then

$$\Omega = \mathcal{R}[(\mathbf{A}\ \mathbf{Y}_1)], \quad \omega = \mathcal{R}[(\mathbf{1}_n\ \mathbf{Y}_1)],$$

dim(Ω) = $q + k$, and dim(ω) = $q + 1$. The likelihood ratio criterion λ can be expressed as

$$\lambda^{2/n} = \Lambda = \frac{|\mathbf{S}_\Omega|}{|\mathbf{S}_\omega|} = \frac{|\mathbf{S}_\Omega|}{|\mathbf{S}_\Omega + (\mathbf{S}_\omega - \mathbf{S}_\Omega)|},$$

where $\mathbf{S}_\Omega = \mathbf{Y}_2'(\mathbf{I}_n - \mathbf{P}_\Omega)\mathbf{Y}_2$ and $\mathbf{S}_\omega = \mathbf{Y}_2'(\mathbf{I}_n - \mathbf{P}_\omega)\mathbf{Y}_2$. We express the LRC in terms of **W** , **B** , and **T**. Let us partition **W**, **B**, and **T** as

$$\mathbf{W} = \begin{pmatrix} \mathbf{W}_{11} & \mathbf{W}_{12} \\ \mathbf{W}_{21} & \mathbf{W}_{22} \end{pmatrix}, \quad \mathbf{B} = \begin{pmatrix} \mathbf{B}_{11} & \mathbf{B}_{12} \\ \mathbf{B}_{21} & \mathbf{B}_{22} \end{pmatrix}, \quad \mathbf{T} = \begin{pmatrix} \mathbf{T}_{11} & \mathbf{T}_{12} \\ \mathbf{T}_{21} & \mathbf{T}_{22} \end{pmatrix},$$

where $\mathbf{W}_{12}: q \times (p-q)$, $\mathbf{B}_{12}: q \times (p-q)$, and $\mathbf{T}_{12}: q \times (p-q)$. Notingthat $\mathbf{P}_\Omega = \mathbf{P}_\mathbf{A} + \mathbf{P}_{(\mathbf{I}_n - \mathbf{P}_\mathbf{A})\mathbf{Y}_1}$, we have

$$\begin{aligned}\mathbf{S}_\Omega &= \mathbf{Y}_2'\{\mathbf{I}_n - \mathbf{P}_\mathbf{A} - (\mathbf{I}_n - \mathbf{P}_\mathbf{A})\mathbf{Y}_1\{\mathbf{Y}_1'(\mathbf{I}_n - \mathbf{P}_\mathbf{A})\mathbf{Y}_1\}^{-1}\mathbf{Y}_1(\mathbf{I}_n - \mathbf{P}_\mathbf{A})\}\mathbf{Y}_2 \\ &= \mathbf{W}_{22} - \mathbf{W}_{21}\mathbf{W}_{11}^{-1}\mathbf{W}_{12} = \mathbf{W}_{22\cdot 1}.\end{aligned}$$

Similarly, noting that $\mathbf{P}_\omega = \mathbf{P}_{\mathbf{1}_n} + \mathbf{P}_{(\mathbf{I}_n - \mathbf{P}_{\mathbf{1}_n})\mathbf{Y}_1}$, we have

$$\begin{aligned}\mathbf{S}_\omega &= \mathbf{Y}_2'\{\mathbf{I}_n - \mathbf{P}_{\mathbf{1}_n} - (\mathbf{I}_n - \mathbf{P}_{\mathbf{1}_n})\mathbf{Y}_1\{\mathbf{Y}_1'(\mathbf{I}_n - \mathbf{P}_{\mathbf{1}_n})\mathbf{Y}_1\}^{-1}\mathbf{Y}_1'(\mathbf{I}_n - \mathbf{P}_{\mathbf{1}_n})\}\mathbf{Y}_2 \\ &= \mathbf{T}_{22} - \mathbf{T}_{21}\mathbf{T}_{11}^{-1}\mathbf{T}_{12} = \mathbf{T}_{22\cdot 1}.\end{aligned}$$

Theorem 6.2 *Suppose that the observation matrix* **Y** *in (6.1) is a set of samples from* $N_p(\theta_i, \Sigma)$, *i = 1, ..., q. Then the likelihood ratio criterion λ for the hypothesis* $H_{2\cdot 1}$ *in (6.8) is given by*

$$\lambda = \left(\frac{|\mathbf{W}_{22\cdot 1}|}{|\mathbf{T}_{22\cdot 1}|}\right)^{n/2},$$

where **W** *and* **T** *are given by (6.6). Further, under* $H_{2\cdot 1}$,

$$\frac{|\mathbf{W}_{22\cdot 1}|}{|\mathbf{T}_{22\cdot 1}|} \sim \Lambda_{p-k}(q-1, n-q-k).$$

Proof. We consider the conditional distributions of $\mathbf{W}_{22\cdot 1}$ and $\mathbf{T}_{22\cdot 1}$ given $\mathbf{Y}_1$ by using Theorem 2.3, and see also that they do not depend on $\mathbf{Y}_1$. We have seen that

$$\mathbf{W}_{22\cdot 1} = \mathbf{Y}_2'\mathbf{Q}_1\mathbf{Y}_2, \quad \mathbf{Q}_1 = \mathbf{I}_n - \mathbf{P}_\mathbf{A} - \mathbf{P}_{(\mathbf{I}_n - \mathbf{P}_\mathbf{A})\mathbf{Y}_1}.$$

It is easy to see that $\mathbf{Q}_1^2 = \mathbf{Q}_1$, $\text{rank}(\mathbf{Q}_1) = \text{tr}\mathbf{Q}_1 = n-q-k$, $\mathbf{Q}_1\mathbf{A} = \mathbf{O}$, $\mathbf{Q}_1\mathbf{X}_1 = \mathbf{O}$, and

$$\mathrm{E}(\mathbf{Y}_2 \mid \mathbf{Y}_1)'\mathbf{Q}_1\mathrm{E}(\mathbf{Y}_2 \mid \mathbf{Y}_1) = \mathbf{O}.$$

This implies that $\mathbf{W}_{22\cdot 1} \mid \mathbf{Y}_1 \sim W_{p-k}(n-q-k, \Sigma_{22\cdot 1})$, and hence $\mathbf{W}_{22\cdot 1} \sim W_{p-k}(n-q-k, \Sigma_{22\cdot 1})$. For $\mathbf{T}_{22\cdot 1}$, we have

$$\mathbf{T}_{22\cdot 1} = \mathbf{Y}_2'\mathbf{Q}_2\mathbf{Y}_2, \quad \mathbf{Q}_2 = \mathbf{I}_n - \mathbf{P}_{1_n} - \mathbf{P}_{(\mathbf{I}_n - \mathbf{P}_{1_n})\mathbf{Y}_1},$$

and hence

$$\mathbf{T}_{22\cdot 1} - \mathbf{W}_{22\cdot 1} = \mathbf{Y}_{2'}(\mathbf{Q}_2 - \mathbf{Q}_1)\mathbf{Y}_2.$$

Similarly, $\mathbf{Q}_2$ is idempotent. Using $\mathbf{P}_{1_n}\mathbf{P}_\mathbf{A} = \mathbf{P}_\mathbf{A}\mathbf{P}_{1_n} = \mathbf{P}_{1_n}$, we have $\mathbf{Q}_1\mathbf{Q}_2 = \mathbf{Q}_2\mathbf{Q}_1 = \mathbf{Q}_1$, and hence

$$(\mathbf{Q}_2 - \mathbf{Q}_1)^2 = \mathbf{Q}_2 - \mathbf{Q}_1, \quad \mathbf{Q}_1 \cdot (\mathbf{Q}_2 - \mathbf{Q}_1) = \mathbf{O}.$$

Further, under $H_{2\cdot 1}$,

$$\mathrm{E}(\mathbf{X}_2 \mid \mathbf{X}_1)'(\mathbf{Q}_2 - \mathbf{Q}_1)\mathrm{E}(\mathbf{X}_2 \mid \mathbf{X}_1) = \mathbf{O}.$$

7. GENERAL MULTIVARIATE LINEAR MODEL

In this section, we consider a general multivariate linear model as follows. Let **Y** be an $n \times p$ observation matrix whose rows are independently distributed as p-variate normal distribution with a common covariance matrix Σ.Σ. Suppose that the mean of **Y** is given as:

$$\mathrm{E}(\mathbf{Y}) = \mathbf{A\Theta X}',$$

where **A** is an $n \times k$ given matrix with rank k, **X** is a $p \times q$ matrix with rank q, and ΘΘ is a $k \times q$ unknown parameter matrix. For a motivation of (7.1), consider the case when a single variable y is measured at p time points $t_1, \ldots, t_p$ (or different conditions) on n subjects chosen at random from a group. Suppose that we denote the variable y at time point t_j by y_j. Let the observations $y_{i1}, \ldots, y_{ip}$ of the ith subject be denoted by

$$\boldsymbol{y}_i = (y_{i1}, \ldots, y_{ip})', \quad i = 1, \ldots, n.$$

If we consider a polynomial regression of degree $q - 1$ of y on the time variable t, then

$$\mathrm{E}(\boldsymbol{y}_i) = \mathbf{X}\boldsymbol{\theta},$$

where

$$\mathbf{X} = \begin{pmatrix} 1 & t_1 & \cdots & t_1^{q-1} \\ \vdots & \vdots & \cdots & \vdots \\ 1 & t_p & \cdots & t_p^{q-1} \end{pmatrix}, \quad \boldsymbol{\theta} = \begin{pmatrix} \theta_1 \\ \theta_2 \\ \vdots \\ \theta_q \end{pmatrix}.$$

If there are k different groups and each group has a polynomial regression of degree $q - 1$ of y, we have a model given by (7.1). From such motivation, the model (7.1) is also called a growth curve model. For its detail, see [14].

Now, let us consider to derive LRC for a general linear hypothesis

$$H_g : \mathbf{C\Theta D} = \mathbf{O},$$

against alternatives $K_g : \mathbf{C\Theta D} \neq \mathbf{O}$. Here, $\mathbf{C}$ is a $c \times k$. given matrix with rank c, and $\mathbf{D}$ is a $q \times d$ given matrix with rank d. This problem was discussed by [15–17]. Here, we obtain LRC by reducing it to the problem of obtaining LRC for a general linear hypothesis in a multivariate linear model. In order to relate the model (7.1) to a multivariate linear model, consider the transformation from $\mathbf{Y}$ to $(\mathbf{UV})$:

$$(\mathbf{U}\ \mathbf{V}) = \mathbf{YG}, \quad \mathbf{G} = (\mathbf{G}_1\ \mathbf{G}_2),$$

where $\mathbf{G}_1 = \mathbf{X}(\mathbf{X}'\mathbf{X})^{-1}$, $\mathbf{G}_2 = \tilde{\mathbf{X}}$, and $\tilde{\mathbf{X}}$ are a $p \times (p - q)$ matrix satisfying $\tilde{\mathbf{X}}'\mathbf{X} = \mathbf{O}$ and $\tilde{\mathbf{X}}'\tilde{\mathbf{X}} = \mathbf{I}_{p-q}$. Then, the rows of $(\mathbf{UV})$ are independently distributed as p-variate normal distributions with means

$$\mathrm{E}[(\mathbf{U}\ \mathbf{V})] = (\mathbf{A\Theta}\ \mathbf{O}),$$

and the common covariance matrix

$$\Psi = \mathbf{G}'\Sigma\mathbf{G} = \begin{pmatrix} \mathbf{G}_1'\boldsymbol{\Sigma}\mathbf{G}_1 & \mathbf{G}_1'\boldsymbol{\Sigma}\mathbf{G}_2 \\ \mathbf{G}_2'\boldsymbol{\Sigma}\mathbf{G}_1 & \mathbf{G}_2\boldsymbol{\Sigma}\mathbf{G}_2 \end{pmatrix} = \begin{pmatrix} \boldsymbol{\Psi}_{11} & \boldsymbol{\Psi}_{12} \\ \boldsymbol{\Psi}_{21} & \boldsymbol{\Psi}_{22} \end{pmatrix}.$$

This transformation can be regarded as one from $\mathbf{y} = (y_1, ..., y_p)'$ to a q-variate main variable $\mathbf{u} = (u_1, ..., u_q)'$ and a $(p - q)$-variate auxiliary variable $\mathbf{v} = (v_1, ..., v_{p-q})'$. The model (7.1) is equivalent to the following joint model of two components:

1. The conditional distribution of $\mathbf{U}$ given $\mathbf{V}$ is

$$\mathbf{U}|\mathbf{V} : \mathrm{N}_{n\times q}(\mathbf{A}^*\boldsymbol{\Xi}, \boldsymbol{\Psi}_{11\cdot 2}).$$

2. The marginal distribution of $\mathbf{V}$ is

$$\mathbf{V} : \mathrm{N}_{n\times(p-q)}(\mathbf{O}, \boldsymbol{\Psi}_{22}),$$

where

$$\mathbf{A}^* = (\mathbf{A}\ \mathbf{V}), \quad \mathbf{\Xi} = \begin{pmatrix} \mathbf{\Theta} \\ \mathbf{\Gamma} \end{pmatrix},$$

$$\mathbf{\Gamma} = \mathbf{\Psi}_{22}^{-1}\mathbf{\Psi}_{21}, \mathbf{\Psi}_{11\cdot 2} = \mathbf{\Psi}_{11} - \mathbf{\Psi}_{12}\mathbf{\Psi}_{22}^{-1}\mathbf{\Psi}_{21}.$$

Before we obtain LRC, first we consider the MLEs in (7.1). Applying a general theory of multivariate linear model to (7.4) and (7.5), the MLEs of ΞΞ , Ψ11·2Ψ11·2 , and Ψ22Ψ22 are given by

$$\hat{\Xi} = (\mathbf{A}^{*\prime}\mathbf{A}^*)^{-1}\mathbf{A}^{*\prime}\mathbf{U}, \quad n\hat{\mathbf{\Psi}}_{11\cdot 2} = \mathbf{U}'(\mathbf{I}_n - \mathbf{P}_{\mathbf{A}^*})\mathbf{U}, \quad n\hat{\mathbf{\Psi}}_{22} = \mathbf{V}'\mathbf{V}.$$

Let

$$\mathbf{S} = \mathbf{Y}'(\mathbf{I}_n - \mathbf{P}_\mathbf{A})\mathbf{Y}, \quad \mathbf{W} = \mathbf{G}'\mathbf{S}\mathbf{G} = (\mathbf{U}\ \mathbf{V})'(\mathbf{I}_n - \mathbf{P}_\mathbf{A})(\mathbf{U}\ \mathbf{V}),$$

and partition **W** as

$$\mathbf{W} = \begin{pmatrix} \mathbf{W}_{11} & \mathbf{W}_{12} \\ \mathbf{W}_{21} & \mathbf{W}_{22} \end{pmatrix}, \quad \mathbf{W}_{12} : q \times (p-q).$$

Theorem 7.1 *For an n × p observation matrix* $\mathbf{Y}$, *assume a general multivariate linear model given by (7.1). Then:*

1. *The MLE* $\hat{\Theta}$ *of* Θ *is given by*

$$\hat{\mathbf{\Theta}} = \mathbf{A}(\mathbf{A}'\mathbf{A})^{-1}\mathbf{A}'\mathbf{Y}\mathbf{S}^{-1}\mathbf{X}(\mathbf{X}'\mathbf{S}^{-1}\mathbf{X})^{-1}.$$

2. *The MLE* $\hat{\Psi}_{11\cdot 2}$ *of* $\Psi_{11\cdot 2}$ *is given by*

$$n\hat{\mathbf{\Psi}}_{11\cdot 2} = \mathbf{W}_{11\cdot 2} = (\mathbf{X}'\mathbf{S}^{-1}\mathbf{X})^{-1}.$$

Proof. The MLE of $\mathbf{\Xi}$ is $\hat{\Xi} = (\mathbf{A}^{*\prime}\mathbf{A}^*)^{-1}\mathbf{A}^{*\prime}\mathbf{U}$. The inverse formula (see (5.5)) gives

$$\mathbf{Q} = (\mathbf{A}^{*\prime}\mathbf{A}^*)^{-1} = \begin{pmatrix} (\mathbf{A}'\mathbf{A})^{-1} & \mathbf{O} \\ \mathbf{O} & \mathbf{O} \end{pmatrix} + \begin{pmatrix} -(\mathbf{A}'\mathbf{A})^{-1}\mathbf{A}'\mathbf{V} \\ \mathbf{I}_{p-q} \end{pmatrix} [\mathbf{V}'(\mathbf{I}_n - \mathbf{P}_\mathbf{A})\mathbf{V}]^{-1} \begin{pmatrix} -(\mathbf{A}'\mathbf{A})^{-1}\mathbf{A}'\mathbf{V} \\ \mathbf{I}_{p-q} \end{pmatrix}'$$
$$= \begin{pmatrix} \mathbf{Q}_{11} & \mathbf{Q}_{12} \\ \mathbf{Q}_{21} & \mathbf{Q}_{22} \end{pmatrix}.$$

Therefore, we have

$$\hat{\mathbf{\Theta}} = (\mathbf{Q}_{11}\mathbf{A}' + \mathbf{Q}_{12}\mathbf{V}')\mathbf{U}$$
$$= (\mathbf{A}'\mathbf{A})^{-1}\mathbf{A}'\mathbf{Y}\mathbf{G}_1 - (\mathbf{A}'\mathbf{A})^{-1}\mathbf{A}'\mathbf{Y}\mathbf{G}_2(\mathbf{G}_2'\mathbf{S}\mathbf{G}_2)^{-1}\mathbf{G}_2'SG_1.$$

Using

$$G_2(G_2'SG_2)^{-1}G_2' = S^{-1} - G_1(G_1'S^{-1}G_1)^{-1}G_1'S^{-1},$$

we obtain 1. For a derivation of 2, let $\mathbf{B}=(\mathbf{I}_n-\mathbf{P}_A)\mathbf{V}$. Then, using $\mathbf{P}_{A*}=\mathbf{P}_A+\mathbf{P}_B$, the first expression of (1) is obtained. Similarly, the second expression of (2) is obtained.

Theorem 7.2 *Let* $\lambda = \Lambda^{n/2}$ *be the LRC for testing the hypothesis* (7.2) *in the generalized multivariate linear model* (7.1). *Then,*

$$\Lambda = |S_e| / |S_e + S_h|,$$

where

$$S_e = \mathbf{D}'(\mathbf{X}S^{-1}\mathbf{X}')^{-1}\mathbf{D}, \quad S_h = (\mathbf{C}\hat{\Theta}\mathbf{D})'(\mathbf{C}R\mathbf{C}')^{-1}\mathbf{C}\hat{\Theta}\mathbf{D}$$

and

$$R = (\mathbf{A}'\mathbf{A})^{-1} + (\mathbf{A}'\mathbf{A})^{-1}\mathbf{A}'\mathbf{Y}S^{-1}\{S - \mathbf{X}'(\mathbf{X}S^{-1}\mathbf{X}')^{-1}\mathbf{X}\}$$
$$\times S^{-1}\mathbf{Y}'\mathbf{A}(\mathbf{A}'\mathbf{A})^{-1}.$$

Here $\hat{\Theta}$ *is given in Theorem* 7.1.1. *Further, the null distribution is* $\Lambda_d(c, n - k - (p - q))$.

Proof. The test of H_g in (7.2) against alternatives K_g is equivalent to testing

$$H_g : \mathbf{C}^*\Xi\mathbf{D} = \mathbf{O}$$

under the conditional model (7.4), where $\mathbf{C}^* = (\mathbf{C} \ \ \mathbf{O})$. Since the distribution of $\mathbf{V}$ does not depend on H_g, the LR test under the conditional model is the LR test under the unconditional model. Using a general result for a general linear hypothesis given in Theorem 4.1, we obtain

$$\Lambda = |\tilde{S}_e| / |\tilde{S}_e + \tilde{S}_h|,$$

where

$$\tilde{S}_e = \mathbf{D}'\mathbf{U}'(\mathbf{I}_n - \mathbf{A}^*(\mathbf{A}^{*\prime}\mathbf{A}^*)^{-1}\mathbf{A}^{*\prime})\mathbf{U}\mathbf{D},$$
$$\tilde{S}_h = (\mathbf{C}\hat{\Xi}\mathbf{D})'(\mathbf{C}^*(\mathbf{A}^{*\prime}\mathbf{A}^*)^{-1}\mathbf{C}^{*\prime})^{-1}\mathbf{C}\hat{\Xi}\mathbf{D}.$$

By reduction similar to those of MLEs, it is seen that $\tilde{S}_e = S_e$ and $\tilde{S}_h = S_h$. This completes the proof.

8. CONCISE SUMMARY

This chapter examines the concept of LRC (Least Residual Covariance) in a multivariate linear model, with a specific emphasis on the significance of projection matrices. The testing problems under consideration comprise hypotheses regarding the selection of variables or the absence of extra knowledge about a group of variables, in addition to a standard linear hypothesis. It is worth mentioning that algebraic approaches are used to obtain various LRCs and their distributions.

We have not engaged in discussions with LRCs regarding the hypothesis of variable selection in canonical correlation analysis, nor have we examined dimensionality in the multivariate linear model. The solutions to these challenges are documented in [3, 18].

Additional test criteria in multivariate analysis include the Lawley-Hotelling trace criterion and the Bartlett-Nanda-Pillai trace criterion. In this chapter, criteria similar to those proposed in [12] can be suggested for the testing challenges discussed.

The likelihood ratio chi-square tests for the absence of any additional information from a group of variables might be valuable in the process of variable selection. One can suggest model selection criteria, such as AIC, as demonstrated in [19].

REFERENCES

1. Anderson, T. W. (2003). An Introduction to Multivariate Statistical Analysis, 3rd ed. John Wiley & Sons, Inc., Hoboken, N.J.
2. Arnold, S. F. (1981). The Theory of Linear Models and Multivariate Analysis, John Wiley & Sons, Inc., Hoboken, N.J.
3. Fujikoshi, Y., Ulyanov, V. V., and Shimizu, R. (2010). Multivariate Statistics: High-Dimensional and Large-Sample Approximations. John Wiley & Sons, Inc., Hobeken, N.J.
4. Muirhead, R. J. (1982). Aspect of Multivariate Statistical Theory, John Wiley & Sons, Inc., New York.
5. Rencher, A. C. (2002). Methods of Multivariate Analysis, 2nd ed. John Wiley & Sons, Inc., New York.
6. Seber, G. A. F. (1984). Multivariate Observations. John Wiley & Sons, Inc., New York.
7. Seber, G. A. F. (2008). A Matrix Handbook for Statisticians. 2nd edn. John Wiley & Sons, Inc., Hobeken, N.J.
8. Siotani, M., Hayakawa, T., and Fujikoshi, Y. (1985). Modern Multivariate Statistical Analysis: A Graduate Course and Handbook, American Sciences Press, Columbus, OH.
9. Harville, D. A. (1997). Matrix Algebra from a Statistician's Perspective. Springer-Verlag, New York.
10. Rao, C. R. (1973). Linear Statistical Inference and Its Applications, 2nd ed.

John Wiley & Sons, Inc., New York.

11. Rao, C. R. (1948). Tests of significance in multivariate analysis. Biometrika. 35, 58–79.
12. Fujikoshi, Y. (1989). Tests for redundancy of some variables in multivariate analysis. In Recent Developments in Statistical Data Analysis and Inference (Y. Dodge, ed.). Elsevier Science Publishers B.V., Amsterdam, 141–163.
13. Rao, C. R. (1970). Inference on discriminant function coefficients. In Essays in Probability and Statistics, (R. C. Bose, ed.), University of North Carolina Press, Chapel Hill, NC, 587–602.
14. Potthoff, R. F. and Roy, S. N. (1964). A generalized multivariate analysis of variance model useful especially for growth curve problems. Biometrika. 51, 313–326.
15. Gleser, L. J. and Olkin, I. (1970). Linear models in multivariate analysis, In Essays in Probability and Statistics, (R.C. Bose, ed.), University of North Carolina Press, Chapel Hill, NC, 267–292.
16. Khatri, C. G. (1966). A note on a MANOVA model applied to problems in growth curve. Ann. Inst. Statist. Math. 18, 75–86.
17. Kshirsagar, A. M. (1995). Growth Curves. Marcel Dekker, Inc.
18. Fujikoshi, Y. (1982). A test for additional information in canonical correlation analysis, Ann. Inst. Statist. Math., 34, 137–144.
19. Akaike, H. (1973). Information theory and an extension of the maximum likelihood principle. In 2nd International Symposium on Information Theory (B.N. Petrov and F. Csáki, eds.). Akadémia Kiado, Budapest, Hungary, 267–281.

INDEX

www.ingramcontent.com/pod-product-compliance
Ingram Content Group UK Ltd.
Pitfield, Milton Keynes, MK11 3LW, UK
UKHW061537190726
13845UKWH00006B/135

9 789349 964556